Contemporary Agriculture

About the Authors

Myself **A.Krishnamoorthi** I did my Bsc agriculture(2012-16) from VIA Agriculutre college Affiliated to TNAU and I did my Msc Agriculture in Plant geneticr esource (2108-20) from TNAU and I am written 10 book chapters 2 research article and 10 review articles I cleared my ICAR SRF Examination with Rank 4 in the Discipline of Economic Botany and Plant Genetic Resources. Now I am pursuing my PhD at the ICAR-National Bureau of Plant Genetic Resourrces IARI Pusa Campus New Delhi.

Vijay Kumar was born and brought up in Muzaffarnagar district, located in the state of Uttar Pradesh. He is currently stationed in ICAR- Sugarcane Breeding Institute, Regional Centre, Karnal, serving as a Technical Officer in the Plant Breeding Division. He obtained a Bachelor of Science degree in Agriculture from SVBP University of Agriculture & Technology, Modipuram, Meerut, Uttar Pradesh. He also accomplished a Master of Science degree in Genetics & Plant Breeding at SHUATS, Naini, Prayagraj, Uttar Pradesh. He has authored 2 articles, 8 research papers, and 3 abstracts. He has participated in numerous national and international seminars, congresses, and conferences. He received the Young Scientist Award at the 6th International Conference on Histolic Innovations and Technological Advance for Sustainable Agriculture.

Dr. Sapna hails from Rewari, a district in Haryana. She is currently employed as a Senior Research Fellow (DUS) in the Crop Improvement Division at ICAR- IIWBR, Karnal. She graduated (B.Sc. biotechnology) from MDU Rohtak, Haryana. She earned M.Sc. and Ph.D. in plant physiology from HAU in Hisar, Haryana. She qualified ICAR (ASRB)-NET in Plant Physiology, CISR Net in Life sciences & Gate in Botany and Biotechnology. In addition to 12 articles, she has authored 14 research and review papers, 5 book chapters, and 15 abstracts. She has attended numerous conferences, seminars, and congresses both nationally and internationally. She has been awarded Acedemic excellance award, best Article Award, Innovative Article Award & Young Scientist Award in 6th International Conference on Histolic Innovations and Technological Advance for Sustainable Agriculture.

Mrs. Ravinder Kaur is a Horticulture Expert, completed her school education from S.G.R.R. Public School, Bhaniyawala; acquired her B.Sc (Agriculture) Degree in 2017 from Uttaranchal University, Uttarakhand and M.Sc. (Horticulture) in 2019 from V.C.S.G. U.U.H.F Bharsar, Uttarakhand with first division. She is working as an Assistant Professor, School of Agriculture, Dev Bhoomi Uttarakhand University, Dehradun, India. She has awarded with "Excellence in Agricultural Research Award", "Young Scientist Award", "Lal Bahadur Shastri Jaivik KrishakAward" and Young Horticulturist Award in different National and international Conferences and has published many research papers, articles, book and book chapters in many national and international peer reviewed journals. Mrs. Ravinder Kaur has attended various National/international Conferences, Seminars, Workshops and Trainings in both Agriculture and related Forestry fields.

Mr. Pankaj Kumar Maurya as Ph.D pursuing in Horticulture (Vegetable Science) From Naini Agricultural Institute, SHUATS, Naini, Prayagraj, U.P. He has completed B.Sc (Ag) from Bihar Agricultural University, Sabour, Bhagalpur, Bihar and M.Sc (Ag) in Horticulture from Dr. Rajendra Prasad Central Agricultural University, Pusa, Samastipur, Bihar. I also qualify ASRB Net in 2021. He has published 12 paper as author and co-author in NAAS rated journal, 17 book chapter and 5 text book as an author. He has attend different training programme, national and international seminar.

Contemporary Agriculture

A. Krishnamoorthi
Vijay Kumar
Sapna
Ravinder Kaur
Pankaj Kumar Maurya

1168, Sector 13, Urban Estate
Karnal-132 001, Haryana
Tel: 91-84470 75807, 18440 41168
Email: contentvibesppa@gmail.com
www.contentvibes.in

Print ISBN: 978-81-97719-26-4

ebook-ISBN: 978-81-97719-25-7

Perface

Understanding the Landscape of Contemporary Agriculture

In an era marked by rapid technological advancements, shifting climate patterns, and growing global populations, agriculture stands at the crossroads of numerous societal challenges and opportunities. This book delves into the multifaceted world of contemporary agriculture, exploring its evolution, current practices, and future prospects.

The Journey from Tradition to Innovation

Agriculture has always been fundamental to human existence. From the dawn of civilization, when early humans transitioned from hunting and gathering to cultivating crops, to the present day, where precision farming and biotechnology are redefining food production, the agricultural sector has continually adapted and transformed. This journey reflects humanity's resilience and ingenuity in the face of changing environmental and social landscapes.

The Green Revolution and Beyond

The mid-20th century Green Revolution marked a pivotal chapter in agricultural history. By introducing high-yield crop varieties, synthetic fertilizers, and advanced irrigation techniques, it significantly boosted food production and staved off widespread hunger. However, it also introduced new challenges such as environmental degradation, loss of biodiversity, and an increased dependency on chemical inputs. This book examines these complex legacies and how they continue to shape agricultural practices today.

Technological Innovation and Sustainable Practices

In recent decades, agriculture has embraced a wave of technological innovations. Precision farming, utilizing GPS and data analytics, enables farmers to optimize field management, enhancing efficiency and sustainability. Drones and satellite imagery provide real-time data, facilitating better decision-making and resource management. Biotechnology offers solutions to improve crop resilience and nutritional value, addressing food security concerns in a changing climate.

Facing Global Challenges

Contemporary agriculture is not without its challenges. Climate change poses significant threats to crop yields and farming practices worldwide. Soil degradation, water scarcity, and pest resistance require innovative approaches and sustainable practices. Additionally, the need to balance food production with environmental conservation has never been more critical.

A Vision for the Future

This book aims to provide a comprehensive understanding of contemporary agriculture, offering insights into the latest advancements and the pressing issues that farmers, policymakers, and researchers face today. By examining the interplay between technology, sustainability, and global food systems, it aspires to inspire solutions that will secure a resilient and prosperous agricultural future for generations to come.

An Invitation to Explore

We invite readers to embark on this exploration of contemporary agriculture, understanding its complexities, celebrating its innovations, and critically examining its challenges. It is through this informed lens that we can appreciate the vital role agriculture plays in our world and envision a future where it continues to thrive in harmony with our planet.

Contents

1

Introduction: The Changing Landscape of Agriculture

Abstract

Agriculture has undergone profound transformations over the past century, driven by technological advancements, economic shifts, and environmental challenges. This article explores the multifaceted evolution of agriculture, emphasizing the integration of modern technology, sustainable practices, and global policy changes. From the advent of precision farming and biotechnology to the impact of climate change and urbanization, contemporary agriculture is characterized by its complexity and adaptability. This examination provides a foundational understanding of the current agricultural landscape, highlighting key trends, challenges, and opportunities for future growth and sustainability.

Introduction

The Changing Landscape of Agriculture

Historical Context and Evolution

Agriculture, one of the oldest human endeavours, has continually adapted to meet the demands of growing populations and changing environments. Historically, agricultural practices were localized and heavily dependent on manual labour and natural conditions. The Industrial Revolution marked a significant turning point, introducing mechanization and chemical inputs, dramatically increasing productivity. However, the 20th and 21st centuries have witnessed even more rapid and transformative changes.

Technological Advancements

1. **Precision Agriculture**

   Precision agriculture represents a revolutionary approach to farming, leveraging data and technology to optimize field-level management. GPS, remote sensing, and IoT devices allow for precise monitoring and control of variables like soil moisture, crop health, and pest presence. This not only increases efficiency but also reduces environmental impact by minimizing the overuse of water, fertilizers, and pesticides.

2. **Biotechnology and Genetic Engineering**

   The introduction of genetically modified organisms (GMOs) has been one of the most contentious yet impactful advancements. GMOs are engineered to exhibit desirable traits such as pest resistance, drought tolerance, and enhanced nutritional content. Biotechnology extends beyond GMOs, encompassing gene editing techniques like CRISPR, which allow for even more precise genetic modifications.

3. **Automation and Robotics**

   Automation is reshaping labour dynamics in agriculture. From autonomous tractors and harvesters to robotic milking machines, these technologies reduce the need for manual labour and increase operational efficiency. Drones are also becoming invaluable for tasks such as aerial spraying and crop monitoring.

Sustainable Farming Practices

1. **Organic Farming**

   Organic farming eschews synthetic chemicals in favour of natural alternatives and emphasizes biodiversity, soil health, and ecological balance. While organic farming can be more labour-intensive and yield lower outputs, it offers benefits such as improved soil fertility and reduced environmental pollution.

2. **Agroecology**

   Agroecology integrates principles of ecology into agricultural practices. It promotes biodiversity, nutrient cycling, and resilience through practices such as crop rotation, intercropping, and the use of cover crops. Agroecology aims to create sustainable and self-sufficient farming systems.

3. **Regenerative Agriculture**

   Regenerative agriculture focuses on restoring and enhancing soil health, sequestering carbon, and increasing biodiversity. Techniques include no-till farming, holistic grazing, and agroforestry. These practices help combat climate change by capturing carbon in the soil and improving the resilience of farming systems to extreme weather events.

Climate Change and Environmental Impact

1. **Impact on Crop Yields**

   Climate change poses significant threats to agricultural productivity. Changes in temperature, precipitation patterns, and the frequency of extreme weather events can adversely affect crop yields. For example, prolonged droughts can reduce water availability, while unseasonal frosts can damage crops.

2. **Adaptation Strategies**

   To mitigate these impacts, farmers are adopting various adaptation strategies. These include developing drought-resistant crop varieties, altering planting schedules, and implementing water-saving irrigation technologies. Governments and research institutions are also investing in climate-resilient agricultural practices and infrastructure.

3. **Carbon Footprint of Agriculture**

   Agriculture is both a contributor to and a victim of climate change. The sector is responsible for a significant portion of global greenhouse gas emissions, primarily through livestock production, deforestation, and the use of synthetic fertilizers. Efforts to reduce agriculture's carbon footprint include promoting sustainable livestock practices, reforestation, and the use of organic fertilizers.

Urban Agriculture

1. **Rise of Urban Farming**

   As urban populations grow, so does the interest in urban agriculture. Urban farming includes practices such as rooftop gardens, vertical farming, and community gardens. These practices not only provide fresh produce to city dwellers but also reduce food miles and enhance urban resilience.

2. **Technological Integration**

   Urban agriculture often leverages advanced technologies like hydroponics, aeroponics, and aquaponics. These soilless farming methods allow for efficient food production in limited spaces and reduce the dependency on traditional agricultural land.

3. **Social and Economic Benefits**

   Urban agriculture can offer numerous social and economic benefits. It promotes local food security, provides educational opportunities, and can create green jobs. Additionally, urban farms and gardens can enhance community cohesion and provide recreational spaces.

Policy and Economic Shifts

1. **Agricultural Policies**

   Government policies play a crucial role in shaping the agricultural landscape. Subsidies, tariffs, and regulations can influence what crops are grown, how land is used, and the adoption of new technologies. Policies promoting sustainable practices and supporting smallholder farmers are essential for a resilient agricultural sector.

2. **Global Trade and Food Security**

   The global food supply chain is increasingly interconnected. Trade policies and international agreements impact the availability and price of agricultural products. Ensuring food security requires balancing trade, local production, and the ability to respond to global market fluctuations.

3. **Economic Pressures**

   Farmers face numerous economic pressures, including fluctuating commodity prices, high input costs, and competition from large agribusinesses. Access to credit, insurance, and financial services is critical for farmers to invest in new technologies and practices.

Future Trends and Opportunities

1. **Innovation and Research**

   Continued investment in agricultural research and innovation is crucial for addressing future challenges. This includes developing new crop varieties, improving soil health, and finding sustainable pest and disease management solutions.

2. **Sustainability and Resilience**

   Building sustainable and resilient agricultural systems is a priority. This involves adopting practices that protect natural resources, reduce emissions, and enhance the ability of farming systems to withstand shocks and stresses.

3. **Global Collaboration**

   Addressing the complex challenges facing agriculture requires global collaboration. International cooperation on research, policy, and trade can help share knowledge, resources, and technologies to ensure a secure and sustainable food future.

Conclusion

The landscape of contemporary agriculture is dynamic and multifaceted. Technological advancements, sustainable practices, and global policy changes are driving the evolution of the sector. While challenges such as climate change and economic pressures persist, there are significant opportunities for innovation and growth. By embracing these changes and working collaboratively, the agricultural community can create a resilient and sustainable future.

This comprehensive overview of the changing landscape of agriculture provides a foundation for understanding the complexities and opportunities within the sector. It highlights the importance of integrating technology, sustainability, and policy to meet the demands of a growing global population while protecting the environment.

2

Sustainable Farming Practices A Necessity for the Future

Abstract

Sustainable farming practices are essential for ensuring food security, environmental protection, and the long-term viability of agricultural systems. As the global population continues to rise, traditional farming methods are increasingly proving to be unsustainable due to their reliance on chemical inputs, depletion of natural resources, and contribution to climate change. This article delves into the various aspects of sustainable farming, including organic agriculture, agroecologist, and regenerative farming practices. It examines the benefits and challenges associated with these approaches, as well as the technological innovations and policy frameworks that support their adoption. The discussion highlights the necessity of transitioning to sustainable practices to safeguard our planet and future food supply.

Sustainable Farming Practices: A Necessity for the Future

Introduction

The concept of sustainable farming encompasses agricultural methods that prioritize environmental health, economic profitability, and social and economic equity. In contrast to conventional farming practices that often deplete resources and harm ecosystems, sustainable farming seeks to maintain and enhance natural resources, reduce pollution, and promote biodiversity. As we face the dual challenges of climate change and a growing global population, the urgency of adopting sustainable farming practices has never been greater.

Historical Context of Sustainable Farming

Historically, many traditional farming methods were inherently sustainable, relying on crop rotation, polycultures, and organic fertilizers. However, the industrialization of agriculture in the 20th century brought about a shift towards monocultures, heavy machinery, synthetic fertilizers, and pesticides. While these changes significantly increased productivity, they also led to soil degradation, water pollution, and loss of biodiversity. The modern sustainable farming movement seeks to integrate the best of traditional practices with contemporary innovations to create a resilient and productive agricultural system.

Key Principles of Sustainable Farming

1. **Conservation of Resources**
   - **Soil Health**: Maintaining soil fertility and structure through organic amendments, cover cropping, and reduced tillage.
   - **Water Management**: Efficient use of water resources through practices such as drip irrigation, rainwater harvesting, and the use of drought-resistant crop varieties.
   - **Energy Efficiency**: Reducing reliance on fossil fuels by adopting renewable energy sources and energy-efficient technologies.
2. **Biodiversity**
   - **Crop Diversity**: Growing a variety of crops to enhance ecosystem resilience and reduce pest and disease pressures.
   - **Agroforestry**: Integrating trees and shrubs into agricultural landscapes to improve biodiversity, soil health, and carbon sequestration.
   - **Wildlife Habitat Preservation**: Creating and maintaining habitats for beneficial insects, birds, and other wildlife.
3. **Integrated Pest Management (IPM)**
   - **Biological Controls**: Using natural predators and parasites to control pest populations.
   - **Cultural Practices**: Crop rotation, intercropping, and selecting pest-resistant varieties to minimize pest outbreaks.
   - **Mechanical and Physical Controls**: Employing traps, barriers, and manual removal of pests.
4. **Economic Viability and Social Equity**
   - **Fair Labour Practices**: Ensuring fair wages and safe working conditions for farm workers.
   - **Community Support**: Supporting local economies through community-supported agriculture (CSA) and farmers' markets.
   - **Access to Resources**: Ensuring equitable access to land, water, and technology for all farmers, including marginalized and smallholder farmers.

Types of Sustainable Farming Practices

1. **Organic Farming**
   - **Definition and Principles**: Organic farming avoids synthetic chemicals and GMOs, focusing on natural inputs and processes to maintain soil fertility and ecological balance.
   - **Practices**: Composting, green manures, biological pest control, and crop rotations.
   - **Benefits and Challenges**: Organic farming can improve soil health and biodiversity, but may face challenges such as lower yields and higher labour costs.
2. **Agroecology**
   - **Definition and Principles**: Agroecology applies ecological principles to agricultural systems, aiming to create sustainable and resilient farming practices.
   - **Practices**: Polyculture, agroforestry, cover cropping, and integrating livestock and crop production.
   - **Benefits and Challenges**: Agroecology can enhance biodiversity and system resilience, but requires extensive knowledge and may have higher initial costs.
3. **Regenerative Agriculture**
   - **Definition and Principles**: Regenerative agriculture focuses on restoring and enhancing soil health, sequestering carbon, and improving ecosystem services.
   - **Practices**: No-till farming, holistic grazing, composting, and diversified cropping systems.
   - **Benefits and Challenges**: Regenerative practices can significantly improve soil health and carbon sequestration, but adoption may require substantial changes to traditional farming methods.

Technological Innovations in Sustainable Farming

1 **Precision Agriculture**
   - **Tools and Techniques**: GPS mapping, remote sensing, and IoT devices to monitor and manage field variability.
   - **Benefits**: Increased efficiency in the use of water, fertilizers, and pesticides, leading to reduced environmental impact.

2. **Biotechnology**
   - **Applications**: Development of drought-resistant and pest-resistant crop varieties through genetic engineering and CRISPR technology. .
   - **Benefits**: Enhanced crop resilience and productivity, reduced need for chemical inputs.
3. **Renewable Energy Integration**
   - **Technologies**: Solar panels, wind turbines, and bioenergy systems for powering farm operations.
   - **Benefits**: Reduced carbon footprint and lower energy costs.
4. **Robotics and Automation**
   - **Applications**: Autonomous tractors, drones for monitoring and spraying, and robotic harvesters.
   - **Benefits**: Increased efficiency and reduced labour costs.

Policy and Economic Support for Sustainable Farming

1. **Government Policies**
   - **Subsidies and Incentives**: Financial support for farmers adopting sustainable practices, such as subsidies for organic farming or tax incentives for renewable energy use.
   - **Regulations**: Policies to limit the use of harmful chemicals and promote conservation practices.
   - **Research and Development**: Funding for research into sustainable farming technologies and practices.
2. **Market-Based Approaches**
   - **Certification and Labelling**: Organic and fair-trade certifications that provide market advantages for sustainably produced goods.
   - **Carbon Credits**: Financial incentives for practices that sequester carbon or reduce greenhouse gas emissions.
3. **International Collaboration**
   - **Global Agreements**: Participation in international agreements focused on sustainable agriculture and climate change mitigation.
   - **Knowledge Sharing**: Collaboration and knowledge exchange between countries and institutions to promote best practices.

Challenges and Barriers to Adoption

1. **Economic Barriers**
   - **Initial Costs**: High upfront costs for transitioning to sustainable practices and technologies.
   - **Market Access**: Difficulty in accessing premium markets that pay higher prices for sustainably produced goods.
2. **Knowledge and Training**
   - **Education and Training**: Need for extensive knowledge and training in sustainable practices, which may be lacking in many farming communities.
   - **Extension Services**: Limited availability of agricultural extension services to support farmers in the transition to sustainable practices.
3. **Policy and Institutional Barriers**
   - **Regulatory Hurdles**: Complex and sometimes contradictory regulations that hinder the adoption of sustainable practices.
   - **Institutional Support**: Lack of strong institutional frameworks to support sustainable agriculture initiatives.

Case Studies and Success Stories

1. **Organic Farming in Europe**
   - **Example**: The success of organic farming in countries like Denmark and Switzerland, where supportive policies and market demand have driven widespread adoption.
   - **Outcomes**: Improved soil health, increased biodiversity, and economic benefits for farmers.
2. **Agroecology in Latin America**
   - **Example**: Agroecological initiatives in Brazil and Cuba that integrate traditional knowledge with modern science to create resilient farming systems.
   - **Outcomes**: Enhanced food security, improved livelihoods, and greater resilience to climate change.
3. **Regenerative Agriculture in the United States**
   - **Example**: The adoption of regenerative practices by farmers in the Midwest to restore soil health and increase carbon sequestration.
   - **Outcomes**: Increased soil fertility, reduced input costs, and potential carbon credits.

Future Directions and Opportunities

1. **Scaling Up Sustainable Practices**
   - **Strategies**: Increasing investment in sustainable agriculture, enhancing policy support, and fostering market demand for sustainably produced goods.
   - **Potential Impact**: Greater adoption of sustainable practices can lead to improved food security, environmental health, and climate resilience.
2. **Innovative Research and Development**
   - **Focus Areas**: Developing new sustainable technologies, improving crop and livestock varieties, and enhancing ecosystem services.
   - **Expected Outcomes**: Continued innovation can provide new solutions to the challenges facing sustainable agriculture.
3. **Global Collaboration and Knowledge Exchange**
   - **Mechanisms**: International partnerships, networks, and platforms for sharing best practices and lessons learned.
   - **Benefits**: Collaborative efforts can accelerate the transition to sustainable agriculture worldwide.

Conclusion

Sustainable farming practices are not just an option but a necessity for the future of agriculture. They offer a pathway to meet the growing global demand for food while preserving natural resources, protecting the environment, and enhancing social equity. The transition to sustainable agriculture requires concerted efforts from farmers, policymakers, researchers, and consumers. By embracing sustainable practices and supporting innovations, we can build a resilient and sustainable food system for future generations.

This comprehensive article underscores the importance of sustainable farming practices in addressing the pressing challenges of our time. It highlights the need for a multifaceted approach that integrates technology, policy, and traditional knowledge to create a sustainable agricultural landscape.

3

Technological Advancements in Modern Agriculture

Abstract

The agricultural sector is undergoing a transformative revolution driven by technological advancements. This article provides an in-depth exploration of how modern technologies are reshaping farming practices, enhancing productivity, and promoting sustainability. Key innovations such as precision agriculture, biotechnology, automation, and digital farming tools are discussed in detail. The article examines the benefits and challenges associated with these technologies, their impact on different aspects of farming, and the future potential of continued technological integration in agriculture. It highlights the critical role that technology plays in meeting the global food demand, addressing climate change, and ensuring economic viability for farmers.

Technological Advancements in Modern Agriculture

Introduction

Agriculture has always been a dynamic field, evolving to meet the needs of growing populations and changing climates. In recent decades, the pace of technological advancement has accelerated, leading to profound changes in how farming is conducted. These innovations are crucial for addressing the challenges of food security, environmental sustainability, and economic efficiency in the 21st century.

Precision Agriculture

1. **Definition and Overview**
    - **Precision agriculture (PA)**: involves the use of technology to monitor and manage variability in crops and soil within a field. It aims to optimize field-level management concerning crop farming.
2. **Tools and Techniques**
    - **Global Positioning Systems (GPS)**: GPS technology is fundamental to precision agriculture, enabling precise mapping and navigation for farming equipment.

- **Remote Sensing**: Satellites and drones equipped with sensors capture detailed images of fields, providing data on crop health, soil conditions, and moisture levels.
- **Variable Rate Technology (VRT)**: VRT allows the application of inputs (fertilizers, pesticides, water) at variable rates across a field, based on data-driven insights.

3. **Benefits**
    - **Increased Efficiency**: Precise application of inputs reduces waste and lowers costs.
    - **Environmental Impact**: Minimizing the use of chemicals and water helps protect the environment.
    - **Enhanced Productivity**: Real-time data enables better decision-making, leading to higher yields and quality.
4. **Challenges**
    - **High Initial Costs**: The technology and equipment required for PA can be expensive.
    - **Data Management**: Handling and interpreting large volumes of data require specialized skills and software.

Biotechnology

1. **Genetic Modification and CRISPR**
    - **Genetically Modified Organisms (GMOs)**: GMOs are plants or animals whose genetic material has been altered to exhibit desirable traits such as pest resistance, herbicide tolerance, or improved nutritional content.
    - **CRISPR Technology**: CRISPR-Cas9 allows for precise editing of genes, facilitating the development of crop varieties that are more resilient to diseases and environmental stresses.
2. **Applications**
    - **Disease and Pest Resistance**: Crops can be engineered to resist specific pests and diseases, reducing the need for chemical pesticides.
    - **Climate Resilience**: Development of drought-tolerant and heat-resistant crop varieties to adapt to changing climate conditions.
    - **Nutritional Enhancement**: Biofortification of crops to increase their nutritional value, such as Golden Rice enriched with Vitamin A.

3. **Benefits**
   - **Improved Crop Yields**: Biotechnology can significantly increase the productivity of crops.
   - **Reduced Chemical Usage**: Pest and disease-resistant crops require fewer chemical inputs.
   - **Enhanced Food Security**: Improved crop resilience ensures stable food supplies in the face of climate change.
4. **Challenges**
   - **Regulatory Hurdles**: GMOs face strict regulatory scrutiny and public scepticism in many regions.
   - **Ethical Concerns**: Genetic modification raises ethical issues related to biodiversity and food safety.

Automation and Robotics

1. **Autonomous Machinery**
   - **Self-Driving Tractors**: Autonomous tractors equipped with GPS and sensors can perform tasks such as ploughing, seeding, and harvesting without human intervention.
   - **Robotic Harvesters**: Robots capable of picking fruits and vegetables, reducing labour costs and addressing labour shortages.
2. **Drones and UAVs**
   - **Surveillance and Monitoring**: Drones provide aerial views of fields, enabling detailed monitoring of crop conditions.
   - **Spraying and Seeding**: Drones can be used for precise application of pesticides and fertilizers, as well as for seeding.
3. **Benefits**
   - **Labour Efficiency**: Automation reduces the need for manual labour, addressing labour shortages and lowering costs.
   - **Precision**: Robots and drones can perform tasks with high accuracy, improving productivity and reducing waste.
   - **Safety**: Automation can reduce human exposure to hazardous farming activities and chemicals.
4. **Challenges**
   - **Cost and Maintenance**: High initial investment and ongoing maintenance costs can be prohibitive for small-scale farmers.

- **Technological Complexity**: Requires training and technical expertise to operate and maintain.

Digital Farming Tools

1. **Farm Management Software**
    - **Data Integration**: Software platforms integrate data from various sources (weather, soil, equipment) to provide comprehensive farm management insights.
    - **Decision Support Systems**: Tools that help farmers make informed decisions on crop planning, input application, and resource management.
2. **Internet of Things (IoT)**
    - **Sensor Networks**: IoT devices, such as soil moisture sensors and climate monitors, provide real-time data on environmental conditions.
    - **Smart Irrigation Systems**: Automated irrigation systems that adjust watering schedules based on sensor data to optimize water usage.
3. **Artificial Intelligence (AI)**
    - **Predictive Analytics**: AI algorithms analyse data to predict crop yields, pest outbreaks, and weather patterns.
    - **Automation and Robotics**: AI enhances the capabilities of autonomous machinery and robotic systems.
4. **Benefits**
    - **Data-Driven Decisions**: Improved decision-making based on accurate, real-time data.
    - **Resource Optimization**: Efficient use of inputs, reducing costs and environmental impact.
    - **Increased Productivity**: Enhanced management practices lead to higher yields and better quality crops.
5. **Challenges**
    - **Data Privacy and Security**: Concerns about the ownership and security of farm data.
    - **Accessibility**: Ensuring that smallholder and resource-limited farmers have access to digital tools and technologies.

Environmental Impact and Sustainability

1. **Sustainable Practices**
   - **Conservation Tillage**: Reducing soil disturbance to enhance soil health and carbon sequestration.
   - **Cover Cropping**: Growing cover crops to improve soil structure, fertility, and biodiversity.
   - **Integrated Pest Management (IPM)**: Combining biological, cultural, and mechanical methods to manage pests sustainably.
2. **Climate Change Mitigation**
   - **Carbon Sequestration**: Practices like agroforestry and regenerative agriculture that capture and store carbon in soils and vegetation.
   - **Reduced Emissions**: Technologies that reduce greenhouse gas emissions from farming activities, such as precision application of fertilizers.
3. **Resource Efficiency**
   - **Water Conservation**: Technologies like drip irrigation and smart irrigation systems that optimize water use.
   - **Nutrient Management**: Precision application of fertilizers to reduce runoff and improve nutrient uptake by plants.
4. **Benefits**
   - **Environmental Protection**: Reduced pollution and improved ecosystem health.
   - **Climate Resilience**: Enhanced ability of farming systems to withstand climate extremes.
   - **Long-Term Viability**: Sustainable practices ensure the long-term productivity and health of agricultural land.
5. **Challenges**
   - **Adoption Barriers**: High initial costs and the need for technical knowledge can hinder the adoption of sustainable practices.
   - **Policy Support**: Requires supportive policies and incentives to promote widespread adoption.

Policy and Economic Considerations

1. **Government Support**
   - **Subsidies and Incentives**: Financial incentives for adopting sustainable technologies and practices.

- **Research and Development**: Funding for research into new agricultural technologies and practices.

2. **Market Dynamics**
   - **Consumer Demand**: Increasing demand for sustainably produced food can drive the adoption of sustainable practices.
   - **Supply Chain Integration**: Integrating sustainable practices throughout the agricultural supply chain to ensure transparency and accountability.
3. **Economic Viability**
   - **Cost-Benefit Analysis**: Assessing the economic benefits of adopting new technologies relative to their costs.
   - **Risk Management**: Tools and strategies to manage the risks associated with technological adoption, such as insurance and financial planning.
4. **Benefits**
   - **Economic Growth**: Technological advancements can boost productivity and profitability in the agricultural sector.
   - **Job Creation**: New technologies can create jobs in tech development, maintenance, and support.
   - **Global Competitiveness**: Adoption of advanced technologies can enhance the global competitiveness of agricultural producers.
5. **Challenges**
   - **Inequality**: Ensuring equitable access to technology for all farmers, including smallholders and those in developing regions.
   - **Regulatory Hurdles**: Navigating complex regulatory environments to implement new technologies.

Future Directions and Opportunities

1. **Innovation and Research**
   - **New Technologies**: Continued development of cutting-edge technologies to address emerging agricultural challenges.
   - **Interdisciplinary Approaches**: Integrating insights from various fields, such as biology, engineering, and data science, to innovate in agriculture.
2. **Global Collaboration**
   - **Knowledge Sharing**: International cooperation to share best practices and technological advancements.

- **Capacity Building**: Supporting developing countries in building the capacity to adopt and benefit from new technologies.

3. **Sustainable Development Goals (SDGs)**
    - **Alignment**: Ensuring that technological advancements in agriculture align with the United Nations' SDGs, particularly those related to food security, climate action, and sustainable production.

4. **Potential Impact**
    - **Food Security**: Technologies can help ensure a stable and sufficient food supply for the growing global population.
    - **Environmental Sustainability**: Adoption of sustainable practices can mitigate environmental impacts and enhance ecosystem health.
    - **Economic Resilience**: Technological advancements can improve the economic resilience of farmers and rural communities.

Conclusion

Technological advancements in modern agriculture are crucial for meeting the challenges of the 21st century. From precision agriculture and biotechnology to automation and digital tools, these innovations offer significant benefits in terms of productivity, sustainability, and economic viability. However, realizing their full potential requires addressing challenges related to cost, accessibility, and regulatory frameworks. By fostering a supportive environment for technological adoption and ensuring equitable access, we can harness these advancements to build a resilient and sustainable agricultural future.

4

The Role of Biotechnology in Crop Production

Abstract

Biotechnology has emerged as a transformative force in agriculture, offering innovative solutions to enhance crop production. This article explores the extensive role of biotechnology in crop production, tracing its historical development and highlighting key technologies such as genetic engineering, marker-assisted selection, and genomics. The applications of these technologies in developing pest-resistant, herbicide-tolerant, nutritionally enhanced, and climate-resilient crops are examined. Benefits such as increased agricultural productivity, environmental sustainability, economic advantages for farmers, and improved food security are discussed. Challenges, including ethical concerns, environmental risks, socioeconomic disparities, and regulatory issues, are also addressed. The article concludes by considering future prospects for biotechnology in crop production, emphasizing the potential of advanced genetic technologies, synthetic biology, and sustainable agricultural practices.

Introduction

Biotechnology, the fusion of biology and technology, has significantly impacted various sectors, particularly agriculture. In crop production, biotechnology has introduced groundbreaking changes that enhance yield, improve resistance to pests and diseases, and ensure food security. This article delves into the multifaceted role of biotechnology in crop production, exploring its historical development, key technologies, applications, benefits, challenges, and future prospects.

Historical Development of Biotechnology in Agriculture

Early Beginnings

The use of biotechnology in agriculture is not entirely new. Traditional breeding methods, which involve selecting plants with desirable traits and cross-breeding them, have been practiced for centuries. However, these methods were time-consuming and limited by the genetic material available within the same or closely related species.

Advent of Modern Biotechnology

The modern era of biotechnology began in the 20th century with the discovery of the structure of DNA by Watson and Crick in 1953. This breakthrough paved the way for genetic engineering, allowing scientists to directly manipulate the genetic makeup of organisms. The development of recombinant DNA technology in the 1970s marked a significant milestone, enabling the creation of genetically modified organisms (GMOs).

Milestones in Crop Biotechnology

1. **1983**: The first transgenic plant, an antibiotic-resistant tobacco plant, was produced.
2. **1994**: The first commercially available genetically engineered crop, the Flavr Savr tomato, was approved for sale in the United States.
3. **1996**: The introduction of herbicide-tolerant soybeans and insect-resistant cotton marked the widespread adoption of GM crops.

Key Technologies in Crop Biotechnology

Genetic Engineering

Genetic engineering involves the direct manipulation of an organism's genes using biotechnology. This can include inserting genes from one organism into another, deleting or "silencing" genes, or altering the function of genes.

1. **Transgenic Technology**: This involves introducing foreign genes into a plant to confer new traits, such as pest resistance or improved nutritional content. For example, Bt corn, which expresses a bacterial toxin that is toxic to certain pests, is widely used.
2. **CRISPR-Cas9**: A more recent technology, CRISPR-Cas9 allows for precise editing of the genome. This technology can be used to knock out undesirable genes or to insert new genes at specific locations, offering greater control over genetic modifications.

Marker-Assisted Selection (MAS)

Marker-assisted selection uses molecular markers linked to desirable traits to aid in the selection of plants during breeding. This accelerates the breeding process and increases the accuracy of selecting for traits such as disease resistance or drought tolerance.

Genomics and Bioinformatics

The study of the entire genome of crops (genomics) and the use of computational tools to analyze genetic data (bioinformatics) have revolutionized crop biotechnology. These fields allow for the identification of genes associated with important traits and facilitate the development of genetically modified crops.

Tissue Culture and Micropropagation

Tissue culture techniques enable the growth of plants from small tissue samples in a controlled environment. This is useful for cloning genetically modified plants, propagating disease-free plants, and conserving rare species.

Applications of Biotechnology in Crop Production

Pest and Disease Resistance

One of the most significant applications of biotechnology in crop production is the development of pest and disease-resistant crops. Bt crops, which express the insecticidal protein from Bacillus thuringiensis, have been highly effective in controlling pests such as the European corn borer and the cotton bollworm. Similarly, genetically modified crops resistant to viruses, fungi, and bacteria have been developed, reducing the need for chemical pesticides and improving yield.

Herbicide Tolerance

Herbicide-tolerant crops have been engineered to survive applications of specific herbicides, allowing farmers to control weeds more effectively without harming the crop. For example, Roundup Ready soybeans are resistant to glyphosate, a broad-spectrum herbicide. This technology simplifies weed management, reduces labor costs, and promotes conservation tillage practices that help preserve soil health.

Improved Nutritional Content

Biotechnology has been used to enhance the nutritional content of crops, addressing malnutrition and improving public health. Golden Rice, engineered to produce beta-carotene, a precursor of vitamin A, is an example of a crop designed to combat vitamin A deficiency. Similarly, biofortified crops with increased levels of iron, zinc, and other essential nutrients are being developed.

Drought and Salinity Tolerance

Climate change poses significant challenges to agriculture, with increased instances of drought and soil salinity threatening crop yields. Biotechnology offers solutions through the development of crops with enhanced tolerance to these stresses. For example, genetically engineered crops with improved water-use efficiency or salt tolerance can maintain productivity under adverse conditions.

Enhanced Yield and Growth

Biotechnological interventions can lead to crops with improved growth rates and higher yields. This can be achieved through various means, such as optimizing photosynthesis, improving nutrient uptake, and increasing resistance to environmental stresses. These advancements contribute to higher productivity and can help meet the growing demand for food.

Benefits of Biotechnology in Crop Production

Increased Agricultural Productivity

Biotechnology has significantly increased agricultural productivity by developing crops that yield more per hectare. This is crucial for feeding a growing global population, which is projected to reach nearly 10 billion by 2050. Higher yields also mean that less land is needed for agriculture, allowing for the conservation of natural habitats.

Environmental Benefits

Biotechnological advancements contribute to environmental sustainability in several ways:

1. **Reduced Chemical Use**: Pest-resistant and herbicide-tolerant crops reduce the need for chemical pesticides and herbicides, lowering the environmental impact of agriculture and decreasing pollution.
2. **Conservation Tillage**: Herbicide-tolerant crops facilitate no-till or reduced-till farming practices, which help prevent soil erosion, improve water retention, and enhance soil health.
3. **Decreased Greenhouse Gas Emissions**: Higher-yielding crops and improved farming practices reduce the carbon footprint of agriculture by minimizing the need for land conversion and decreasing energy use.

Economic Benefits for Farmers

Farmers benefit economically from biotechnology through higher yields, reduced input costs, and improved crop quality. Pest-resistant and herbicide-tolerant crops lower the expenses associated with pest and weed control, while increased productivity boosts farmers' incomes. Additionally, crops with enhanced nutritional content can fetch higher market prices.

Food Security

Biotechnology plays a crucial role in enhancing food security by ensuring a stable and sufficient food supply. Improved crop yields, resistance to pests and diseases, and resilience to climate change contribute to a reliable production system that can meet global food demands.

Challenges and Concerns in Crop Biotechnology

Ethical and Safety Concerns

The use of genetically modified organisms (GMOs) in agriculture has raised ethical and safety concerns. Critics argue that GMOs may have unforeseen health effects on humans and animals. Although extensive testing and regulatory frameworks are in place to ensure the safety of GM crops, public skepticism remains.

Environmental Risks

There are concerns about the potential environmental risks of GM crops, such as:

1. **Gene Flow**: The transfer of genes from GM crops to wild relatives or non-GM crops through cross-pollination can have unintended ecological consequences.
2. **Resistance Development**: Pests and weeds may develop resistance to genetically engineered traits, such as Bt toxins or herbicides, rendering these technologies less effective over time.
3. **Biodiversity Loss**: The widespread adoption of a few genetically modified crop varieties could reduce genetic diversity, making agricultural systems more vulnerable to diseases and changing environmental conditions.

Socioeconomic Issues

The adoption of biotechnology in agriculture can exacerbate socioeconomic disparities. Smallholder farmers may struggle to afford the technology or access the benefits of GM crops. Additionally, the consolidation of seed markets by a few multinational corporations raises concerns about market monopolies and the control of the food supply.

Regulatory and Intellectual Property Issues

The regulation of GM crops varies by country, creating a complex landscape for their development and commercialization. Intellectual property rights and patent issues also pose challenges, as companies seek to protect their biotechnological innovations while ensuring access for farmers.

Future Prospects of Biotechnology in Crop Production

Advanced Genetic Technologies

Emerging genetic technologies, such as gene editing with CRISPR-Cas9, offer promising prospects for the future of crop biotechnology. These technologies provide greater precision and efficiency in modifying crop genomes, enabling the development of crops with multiple beneficial traits.

Synthetic Biology

Synthetic biology involves designing and constructing new biological parts, devices, and systems. In agriculture, synthetic biology can be used to create entirely new metabolic pathways in plants, leading to the production of valuable compounds or improved resistance to environmental stresses.

Sustainable Agriculture

Biotechnology will continue to play a key role in promoting sustainable agriculture. Innovations aimed at enhancing soil health, reducing greenhouse gas emissions, and improving water-use efficiency will contribute to the sustainability of agricultural systems.

Global Food Security

As the global population grows, biotechnology will be essential in ensuring food security. Developing crops that can thrive in diverse environmental conditions, including those impacted by climate change, will be critical in meeting future food demands.

Public Engagement and Acceptance

For biotechnology to reach its full potential in crop production, public engagement and acceptance are crucial. Transparent communication about the benefits, risks, and safety of GM crops can help build public trust and support for biotechnological innovations.

Conclusion

Biotechnology has transformed crop production, offering solutions to many of the challenges facing modern agriculture. From improving yields and nutritional content to enhancing resistance to pests and environmental stresses, biotechnological advancements have far-reaching benefits. However, addressing the ethical, environmental, and socioeconomic concerns associated with GM crops is essential to ensuring the responsible and equitable use of biotechnology in agriculture. As new technologies emerge and our understanding of plant genetics deepens, biotechnology will continue to play a pivotal role in securing a sustainable and food-secure future.

References

Abah, J., Ishaq, M. N. and Wada, A. C. (2010). The role of biotechnology in ensuring food security and sustainable agriculture. African Journal of Biotechnology, 9(52), 8896-8900.

Clarke, J. L. and Daniell, H. (2011). Plastid biotechnology for crop production: present status and future perspectives. Plant molecular biology, 76, 211-220.

Das, S., Ray, M. K., Panday, D. and Mishra, P. K. (2023). Role of biotechnology in creating sustainable agriculture. PLOS Sustainability and Transformation, 2(7), e0000069.

Estrada, A. C., Díaz, D. V. and Hernández, C. A. M. (2017). The role of biotechnology in agricultural production and food supply. Cienciae investigación agraria: revista latinoamericana de ciencias de la agricultura, 44(1), 1-11.

Maryam, B. M., Datsugwai, M. S. S. and Shehu, I. (2017). The role of biotechnology in food production and processing. Industrial engineering, 1(1), 24-35.

Varshney, R. K., Bansal, K. C., Aggarwal, P. K., Datta, S. K. and Craufurd, P. Q. (2011). Agricultural biotechnology for crop improvement in a variable climate: hope or hype?. Trends in plant science, 16(7), 363-371.

5

Precision Agriculture Farming with Data

Abstract

Precision agriculture represents a paradigm shift in farming, leveraging advanced technologies and data analytics to optimize agricultural practices. This article explores the comprehensive role of precision agriculture in modern farming, tracing its evolution and examining key technologies such as GPS, GIS, remote sensing, and IoT. The applications of these technologies in soil management, crop monitoring, irrigation, pest control, and yield prediction are discussed. Benefits, including increased productivity, resource efficiency, and environmental sustainability, are highlighted alongside challenges such as high costs, data management complexities, and the digital divide. The article concludes by considering future trends in precision agriculture, emphasizing the potential of artificial intelligence, machine learning, and blockchain in further revolutionizing farming practices.

Introduction

Agriculture has always been at the core of human civilization, evolving from primitive practices to highly sophisticated techniques. Precision agriculture, also known as precision farming or site-specific crop management (SSCM), is the latest innovation that integrates advanced technologies and data analytics to enhance farming efficiency and productivity. By focusing on the specific needs of different areas within a field, precision agriculture aims to optimize inputs, reduce waste, and increase yields.

Historical Development of Precision Agriculture

Early Concepts and Mechanization

The roots of precision agriculture can be traced back to the advent of agricultural mechanization in the 19th and early 20th centuries. The introduction of tractors and mechanical harvesters significantly boosted productivity. However, the idea of site-specific management began to take shape only in the latter half of the 20th century with advancements in technology.

Emergence of Precision Agriculture

The 1980s and 1990s saw the emergence of precision agriculture as a recognized field. The development of the Global Positioning System (GPS) and Geographic Information Systems (GIS) provided the foundation for site-specific management. Farmers could now accurately map fields and monitor various parameters, leading to more informed decision-making.

Technological Advancements

The turn of the 21st century witnessed rapid technological advancements that further propelled precision agriculture. Remote sensing technologies, including satellite imagery and drones, along with the Internet of Things (IoT), revolutionized data collection and analysis. These technologies enabled real-time monitoring and precise management of agricultural activities.

Key Technologies in Precision Agriculture

Global Positioning System (GPS)

GPS technology is fundamental to precision agriculture, providing accurate location data for field mapping and machinery guidance. GPS allows farmers to apply inputs like fertilizers and pesticides with pinpoint accuracy, reducing waste and ensuring even application across the field.

Geographic Information Systems (GIS)

GIS is a powerful tool for managing and analyzing spatial data. In agriculture, GIS helps in creating detailed maps of fields that include information on soil types, crop conditions, and topography. These maps are essential for planning and implementing site-specific management practices.

Remote Sensing

Remote sensing technologies, including satellite imagery, drones, and aerial photography, provide critical data on crop health, soil conditions, and weather patterns. These technologies enable farmers to monitor large areas efficiently and detect issues such as nutrient deficiencies, pest infestations, and water stress early on.

Internet of Things (IoT)

The IoT connects various sensors and devices used in agriculture, creating a network that collects and shares data in real-time. IoT applications in precision agriculture include soil moisture sensors, weather stations, and automated irrigation systems. This interconnected network allows for continuous monitoring and automated decision-making processes.

Data Analytics and Machine Learning

Data analytics and machine learning play a crucial role in processing the vast amounts of data generated by precision agriculture technologies. Advanced algorithms analyze this data to provide actionable insights, predict outcomes, and optimize farming practices. Machine learning models can improve over time, offering increasingly accurate recommendations.

Applications of Precision Agriculture

Soil Management

Precision agriculture enhances soil management by providing detailed information about soil properties across different field zones. Technologies like soil sensors and GIS mapping allow for precise soil sampling, enabling targeted application of fertilizers and amendments. This approach ensures optimal nutrient levels and improves soil health over time.

Crop Monitoring and Health Assessment

Remote sensing and IoT devices enable continuous monitoring of crop health. Multispectral and hyperspectral imaging from drones and satellites can detect stress signs in plants that are not visible to the naked eye. This early detection allows farmers to address issues promptly, preventing yield losses and improving crop quality.

Precision Irrigation

Water management is critical in agriculture, especially in regions with limited water resources. Precision irrigation systems use soil moisture sensors and weather data to determine the exact water needs of crops. Automated irrigation systems then deliver the right amount of water at the right time, conserving water and reducing runoff.

Pest and Disease Management

Precision agriculture technologies help in early detection and management of pests and diseases. Remote sensing can identify affected areas, while data analytics predict pest outbreaks based on environmental conditions. Targeted application of pesticides minimizes chemical use and reduces environmental impact.

Yield Prediction and Mapping

Accurate yield prediction is essential for planning and resource allocation. Precision agriculture uses historical data, remote sensing, and machine learning models to forecast yields. Yield maps generated post-harvest provide insights into field performance, helping farmers make informed decisions for future planting and management.

Benefits of Precision Agriculture

Increased Productivity

Precision agriculture optimizes input use, leading to higher crop yields and better quality produce. By addressing the specific needs of different field zones, farmers can ensure that crops receive the right amount of nutrients, water, and protection.

Resource Efficiency

One of the primary benefits of precision agriculture is the efficient use of resources. Precision application of fertilizers, pesticides, and water reduces waste and lowers production costs. This efficiency also minimizes the environmental impact of farming practices.

Environmental Sustainability

Precision agriculture contributes to environmental sustainability by reducing chemical runoff, conserving water, and promoting soil health. Site-specific management practices help maintain ecological balance and protect natural resources.

Economic Benefits for Farmers

Farmers benefit economically from precision agriculture through increased yields, reduced input costs, and improved crop quality. The data-driven approach also provides greater predictability, allowing farmers to plan better and reduce risks associated with farming.

Enhanced Food Security

By improving productivity and resource efficiency, precision agriculture plays a crucial role in enhancing food security. The ability to produce more food with fewer resources ensures a stable food supply to meet the growing global population's needs.

Challenges and Concerns in Precision Agriculture

High Costs

The initial investment in precision agriculture technologies can be high, posing a barrier for small and medium-sized farms. The cost of equipment, software, and training can be prohibitive, limiting the adoption of these technologies.

Data Management and Integration

Precision agriculture generates vast amounts of data that need to be collected, stored, and analyzed. Managing and integrating this data from various sources can be complex and requires sophisticated software and technical expertise.

Digital Divide

The digital divide between developed and developing regions impacts the adoption of precision agriculture. Access to reliable internet, advanced technologies, and technical support is limited in many rural areas, hindering the implementation of precision farming practices.

Technical Skills and Training

Farmers need technical skills and training to effectively use precision agriculture technologies. Continuous education and support are essential to ensure that farmers can harness the full potential of these innovations.

Privacy and Data Security

The use of digital technologies in agriculture raises concerns about data privacy and security. Farmers must trust that their data will be used responsibly and that adequate measures are in place to protect sensitive information from unauthorized access.

Future Trends in Precision Agriculture

Artificial Intelligence and Machine Learning

The integration of artificial intelligence (AI) and machine learning (ML) will further enhance the capabilities of precision agriculture. AI and ML can analyze complex datasets, identify patterns, and provide real-time recommendations. These technologies will improve decision-making and enable more precise management of agricultural practices.

Blockchain Technology

Blockchain technology offers potential solutions for transparency and traceability in agriculture. By creating immutable records of agricultural transactions and processes, blockchain can enhance supply chain management, improve food safety, and build consumer trust.

Advanced Robotics and Automation

The development of advanced robotics and automation will revolutionize precision agriculture. Autonomous tractors, drones, and robotic harvesters can perform tasks with high precision and efficiency, reducing labor costs and increasing productivity.

Integration of Biotechnology and Precision Agriculture

The integration of biotechnology with precision agriculture will lead to the development of crops with enhanced traits, such as pest resistance and drought tolerance. Precision agriculture can provide the precise environmental conditions required to maximize the potential of these genetically engineered crops.

Sustainable Practices and Climate Resilience

Future trends in precision agriculture will focus on promoting sustainable practices and enhancing climate resilience. Technologies that improve soil health, reduce greenhouse gas emissions, and increase water-use efficiency will be critical in addressing the challenges posed by climate change.

Enhanced Connectivity and 5G Networks

The deployment of 5G networks will enhance connectivity in rural areas, enabling real-time data transmission and communication between devices. Improved connectivity will facilitate the widespread adoption of precision agriculture technologies and support more efficient farming practices.

Conclusion

Precision agriculture represents a significant advancement in farming, leveraging data and technology to optimize agricultural practices. By enhancing productivity, resource efficiency, and environmental sustainability, precision agriculture addresses many challenges facing modern agriculture. However, overcoming barriers such as high costs, data management complexities, and the digital divide is essential for widespread adoption. Future trends, including AI, blockchain, and advanced robotics, hold promise for further revolutionizing precision agriculture. As these technologies continue to evolve, precision agriculture will play a pivotal role in ensuring a sustainable and food-secure future.

References

Bendre, M. R., Thool, R. C., & Thool, V. R. (2015, September). Big data in precision agriculture: Weather forecasting for future farming. In 2015 1st international conference on next generation computing technologies (NGCT) (pp. 744-750). IEEE.

Bhat, S. A., & Huang, N. F. (2021). Big data and ai revolution in precision agriculture: Survey and challenges. Ieee Access, 9, 110209-110222.

Monteiro, A., Santos, S., & Gonçalves, P. (2021). Precision agriculture for crop and livestock farming—Brief review. Animals, 11(8), 2345.

Linaza, M. T., Posada, J., Bund, J., Eisert, P., Quartulli, M., Döllner, J., ... & Lucat, L. (2021). Data-driven artificial intelligence applications for sustainable precision agriculture. Agronomy, 11(6), 1227

Rodríguez, M. A., Cuenca, L., & Ortiz, Á. (2019). Big data transformation in agriculture: From precision agriculture towards smart farming. In Collaborative Networks and Digital Transformation: 20th IFIP WG 5.5 Working Conference on Virtual Enterprises, PRO-VE 2019, Turin, Italy, September 23–25, 2019, Proceedings 20 (pp. 467-474). Springer International Publishing.

Saiz-Rubio, V., & Rovira-Más, F. (2020). From smart farming towards agriculture 5.0: A review on crop data management. Agronomy, 10(2), 207.

6

Climate Change and Its Impact on Agriculture

Abstract

Climate change is exerting profound effects on global agriculture, posing significant challenges to food security, agricultural productivity, and rural livelihoods. This article explores the intricate relationship between climate change and agriculture, examining how changes in temperature, precipitation patterns, and extreme weather events impact crop yields, livestock, and farming practices. It delves into the economic, social, and environmental repercussions of these changes, highlighting vulnerable regions and populations. The article also discusses adaptation and mitigation strategies, emphasizing the roles of technology, policy, and sustainable agricultural practices in building resilience. By understanding the multifaceted impact of climate change on agriculture, we can develop comprehensive approaches to ensure food security and sustainable development in an increasingly unpredictable climate.

Introduction

Agriculture is both a contributor to and a victim of climate change. The sector is responsible for a significant share of greenhouse gas emissions, while simultaneously being highly sensitive to climate variations. Climate change affects agriculture in multiple ways, from altering growing seasons and crop productivity to increasing the frequency of extreme weather events. Understanding these impacts is crucial for developing strategies to mitigate risks and ensure food security.

Historical Context

Pre-Industrial Era to Present

The relationship between climate and agriculture has been evident throughout history. Agricultural practices have always been influenced by climatic conditions, with societies developing farming techniques suited to their local environments. However, the industrial era marked a turning point, as fossil fuel use and deforestation led to increased greenhouse gas emissions, exacerbating climate change.

Modern Climate Change and Agriculture

The late 20th and early 21st centuries have seen unprecedented changes in global climate patterns, with significant implications for agriculture. Rising temperatures, shifting precipitation patterns, and an increase in extreme weather events have created new challenges for farmers worldwide. These changes necessitate a re-evaluation of traditional farming practices and the development of new strategies to cope with climate variability.

Key Impacts of Climate Change on Agriculture

Temperature Changes

Increased Average Temperatures

Rising global temperatures have a direct impact on crop growth and productivity. Most crops have optimal temperature ranges, and deviations from these ranges can reduce yields. Higher temperatures can accelerate crop maturation, reducing the time available for biomass accumulation and leading to lower yields.

Heat Stress

Heat stress affects both plants and animals. For crops, extreme heat can damage cellular structures, inhibit photosynthesis, and lead to wilting. For livestock, heat stress can reduce feed intake, milk production, and reproductive performance, increasing mortality rates in severe cases.

Precipitation Patterns

Changes in Rainfall Distribution

Climate change is altering rainfall patterns, leading to both increases and decreases in precipitation in different regions. These changes affect water availability for irrigation and can lead to more frequent and severe droughts and floods, impacting crop yields and soil health.

Droughts and Water Scarcity

Extended periods of drought can severely reduce crop yields and lead to the depletion of water resources. Water scarcity not only affects crop irrigation but also has implications for livestock, reducing water availability for drinking and forage production.

Extreme Weather Events

Increased Frequency and Intensity

Extreme weather events, such as hurricanes, floods, and hailstorms, are becoming more frequent and severe due to climate change. These events can cause immediate damage to crops and infrastructure, as well as long-term impacts on soil quality and agricultural productivity.

Impact on Soil Health

Flooding can lead to soil erosion, nutrient loss, and decreased soil fertility, while droughts can lead to soil degradation and desertification. Both extremes impact the long-term sustainability of agricultural lands.

Pests and Diseases

Shifts in Pest and Disease Patterns

Climate change affects the distribution and prevalence of agricultural pests and diseases. Warmer temperatures and altered precipitation patterns create favorable conditions for certain pests and pathogens, leading to increased crop damage and loss.

Increased Pesticide Use

To combat the rise in pests and diseases, farmers may resort to increased pesticide use, which can have environmental and health consequences. Additionally, pests may develop resistance to pesticides, exacerbating the problem.

Economic, Social, and Environmental Impacts

Economic Impacts

Reduced Crop Yields and Income

Climate change can lead to reduced crop yields, affecting farmers' incomes and food supply. Lower yields can increase food prices, impacting consumers and leading to economic instability in agricultural communities.

Increased Production Costs

Farmers may face higher production costs due to the need for additional inputs such as water, fertilizers, and pesticides. Investment in new technologies and infrastructure to cope with climate change also adds to the financial burden.

Social Impacts

Food Security and Nutrition

Climate change threatens food security by reducing the availability, accessibility, and quality of food. Declining crop yields and increased food prices can lead to malnutrition and food shortages, particularly in vulnerable regions.

Rural Livelihoods and Migration

The adverse effects of climate change on agriculture can undermine rural livelihoods, leading to increased poverty and migration. Displaced rural populations often move to urban areas, straining infrastructure and resources.

Environmental Impacts

Biodiversity Loss

Changes in climate and agricultural practices can lead to biodiversity loss. The shift towards monoculture and the use of chemical inputs can harm ecosystems, reducing biodiversity and resilience to environmental changes.

Greenhouse Gas Emissions

Agriculture contributes to greenhouse gas emissions through activities such as deforestation, rice production, and livestock farming. These emissions exacerbate climate change, creating a feedback loop that further impacts agriculture.

Regional Vulnerabilities

Developing Countries

Sub-Saharan Africa

Sub-Saharan Africa is highly vulnerable to climate change due to its reliance on rain-fed agriculture and limited adaptive capacity. Frequent droughts and changing precipitation patterns threaten food security and livelihoods in the region.

South Asia

South Asia faces significant challenges from climate change, including extreme heat, variable monsoons, and sea-level rise. These factors impact agricultural productivity and increase the risk of food insecurity.

Developed Countries

North America

In North America, climate change affects agriculture through increased temperatures, altered precipitation patterns, and more frequent extreme weather events. These changes impact crop yields and water availability, particularly in the western United States.

Europe

Europe's agriculture is vulnerable to climate change due to heatwaves, droughts, and flooding. Southern Europe is particularly at risk, with potential declines in crop yields and water availability.

Adaptation Strategies

Technological Innovations

Climate-Resilient Crops

The development of climate-resilient crop varieties through traditional breeding and biotechnology can help mitigate the impacts of climate change. These crops are designed to withstand heat, drought, and pests, ensuring stable yields under changing conditions.

Precision Agriculture

Precision agriculture technologies, such as GPS-guided equipment and remote sensing, enable farmers to optimize resource use and improve crop management. These technologies can enhance efficiency and reduce vulnerability to climate variability.

Sustainable Agricultural Practices

Conservation Agriculture

Conservation agriculture practices, including no-till farming, crop rotation, and cover cropping, can improve soil health and resilience to climate change. These practices reduce soil erosion, enhance water retention, and increase biodiversity.

Agroforestry

Agroforestry integrates trees and shrubs into agricultural systems, providing multiple benefits such as improved soil fertility, carbon sequestration, and biodiversity conservation. This approach can enhance the resilience of farming systems to climate change.

Policy and Institutional Support

Climate-Smart Agriculture Policies

Governments and institutions can support climate adaptation through policies that promote climate-smart agriculture. These policies should focus on research and development, extension services, and financial support for farmers adopting sustainable practices.

Insurance and Risk Management

Agricultural insurance and risk management tools can help farmers cope with the uncertainties of climate change. Crop insurance, weather-indexed insurance, and other financial instruments provide a safety net, reducing the economic impact of climate-related losses.

Mitigation Strategies

Reducing Emissions from Agriculture

Sustainable Livestock Management

Improving livestock management practices can reduce greenhouse gas emissions from the agriculture sector. Strategies include optimizing feed efficiency, improving manure management, and promoting rotational grazing.

Rice Production Techniques

Modifying rice production techniques, such as alternate wetting and drying (AWD), can reduce methane emissions. AWD involves periodically draining rice fields, reducing the anaerobic conditions that produce methane.

Carbon Sequestration

Soil Carbon Sequestration

Soil carbon sequestration practices, such as cover cropping and reduced tillage, enhance the storage of carbon in soils. These practices improve soil health and contribute to climate change mitigation.

Reforestation and Afforestation

Reforestation and afforestation projects can sequester carbon while providing additional benefits such as biodiversity conservation and watershed protection. Integrating these projects into agricultural landscapes can enhance their effectiveness.

Renewable Energy Integration

On-Farm Renewable Energy

Integrating renewable energy sources, such as solar and wind, into agricultural operations can reduce reliance on fossil fuels and lower greenhouse gas emissions. On-farm renewable energy systems can also provide additional income streams for farmers.

Bioenergy Crops

Growing bioenergy crops, such as switchgrass and miscanthus, can provide renewable energy while also contributing to carbon sequestration. These crops can be integrated into existing agricultural systems, enhancing sustainability.

Case Studies

Sub-Saharan Africa: Drought-Resilient Crops

In Sub-Saharan Africa, the development and adoption of drought-resilient crop varieties have helped mitigate the impacts of climate change. Projects focusing on drought-tolerant maize and millet have improved food security and livelihoods in drought-prone regions.

South Asia: Integrated Water Management

In South Asia, integrated water management strategies have been implemented to address water scarcity and improve agricultural productivity. Techniques such as rainwater harvesting, drip irrigation, and efficient water use have enhanced resilience to changing precipitation patterns.

North America: Precision Agriculture Adoption

North American farmers have widely adopted precision agriculture technologies to optimize resource use and improve crop management. GPS-guided machinery, remote sensing, and variable-rate technology have increased efficiency and reduced vulnerability to climate variability.

Europe: Agroforestry Practices

European farmers are increasingly integrating agroforestry practices into their agricultural systems. These practices enhance biodiversity, improve soil health, and provide additional income streams, contributing to resilience against climate change.

Future Directions

Research and Innovation

Climate-Adapted Crop Breeding

Continued research and innovation in crop breeding are essential for developing climate-adapted varieties. Advances in biotechnology and genomics can accelerate the development of crops that are resilient to heat, drought, and pests.

Digital Agriculture

The adoption of digital agriculture technologies, such as artificial intelligence, machine learning, and big data analytics, can revolutionize farming practices. These technologies enable precise monitoring, predictive analytics, and informed decision-making, enhancing resilience to climate change.

Global Cooperation

International Climate Agreements

Global cooperation through international climate agreements, such as the Paris Agreement, is crucial for addressing the impacts of climate change on agriculture. These agreements provide frameworks for reducing emissions, promoting adaptation, and supporting sustainable development.

Knowledge Sharing and Capacity Building

Knowledge sharing and capacity building are essential for disseminating best practices and technologies. International organizations, governments, and non-governmental organizations can facilitate the exchange of knowledge and resources, empowering farmers to adapt to climate change.

Policy and Governance

Supportive Policies and Incentives

Governments must implement supportive policies and incentives to encourage the adoption of climate-smart agriculture practices. Financial support, subsidies, and regulatory frameworks can promote sustainable farming and enhance resilience.

Multi-Stakeholder Engagement

Engaging multiple stakeholders, including farmers, researchers, policymakers, and the private sector, is crucial for developing comprehensive solutions to climate change. Collaborative approaches ensure that diverse perspectives and expertise are integrated into adaptation and mitigation strategies.

Conclusion

Climate change poses significant challenges to agriculture, impacting crop yields, livestock, and farming practices. The economic, social, and environmental repercussions of these changes are profound, particularly for vulnerable regions and populations. However, through technological innovations, sustainable agricultural practices, and supportive policies, it is possible to build resilience and ensure food security in the face of climate change. By fostering global cooperation and investing in research and capacity building, we can develop comprehensive strategies to mitigate the impacts of climate change on agriculture and create a sustainable future for all

References

Bendre, M. R., Thool, R. C., & Thool, V. R. (2015, September). Big data in precision agriculture: Weather forecasting for future farming. In 2015 1st international conference on next generation computing technologies (NGCT) (pp. 744-750). IEEE.

Bhat, S. A., & Huang, N. F. (2021). Big data and ai revolution in precision agriculture: Survey and challenges. Ieee Access, 9, 110209-110222.

Linaza, M. T., Posada, J., Bund, J., Eisert, P., Quartulli, M., Döllner, J., ... & Lucat, L. (2021). Data-driven artificial intelligence applications for sustainable precision agriculture. Agronomy, 11(6), 1227

Monteiro, A., Santos, S., & Gonçalves, P. (2021). Precision agriculture for crop and livestock farming—Brief review. Animals, 11(8), 2345.

Rodríguez, M. A., Cuenca, L., & Ortiz, Á. (2019). Big data transformation in agriculture: From precision agriculture towards smart farming. In Collaborative Networks and Digital Transformation: 20th IFIP WG 5.5 Working Conference on Virtual Enterprises, PRO-VE 2019, Turin, Italy, September 23–25, 2019, Proceedings 20 (pp. 467-474). Springer International Publishing.

Saiz-Rubio, V., & Rovira-Más, F. (2020). From smart farming towards agriculture 5.0: A review on crop data management. Agronomy, 10(2), 207.

7

Urban Agriculture Bringing Farming to the City

Abstract

Urban agriculture is gaining prominence as a viable solution to numerous challenges faced by modern cities, including food security, environmental sustainability, and community well-being. This comprehensive article explores the concept of urban agriculture, its historical development, and its various forms and practices. It examines the benefits of urban farming, such as enhanced food security, reduced food miles, and improved urban environments, while also addressing challenges like space constraints and regulatory issues. Case studies from cities worldwide illustrate the successful implementation of urban agriculture. The article concludes by discussing future trends and the potential of urban agriculture to transform urban landscapes into productive, sustainable, and resilient communities.

Introduction

As the global urban population continues to grow, cities face increasing challenges related to food security, environmental sustainability, and community health. Urban agriculture, the practice of cultivating, processing, and distributing food in or around urban areas, has emerged as a promising solution to these challenges. By integrating food production into urban landscapes, cities can enhance food security, reduce their ecological footprint, and foster stronger communities.

Historical Development of Urban Agriculture

Ancient and Medieval Cities

Urban agriculture is not a new concept; it has been practiced for thousands of years. In ancient civilizations such as Mesopotamia, Egypt, and the Indus Valley, cities incorporated agricultural activities within their boundaries to ensure food supply. Medieval European cities also featured gardens and small farms within city walls.

Industrial Revolution to the 20th Century

The Industrial Revolution brought significant changes to urban living, with rapid urbanization and the separation of food production from urban areas. However, during times of crisis, such as World War I and World War II, urban agriculture resurfaced in the form of victory gardens, where urban residents grew food to supplement rationed supplies.

Modern Revival

In recent decades, urban agriculture has experienced a revival, driven by concerns about food security, sustainability, and community resilience. This resurgence is characterized by diverse forms of urban farming, from community gardens and rooftop farms to vertical agriculture and aquaponics.

Forms and Practices of Urban Agriculture

Community Gardens

Community gardens are shared spaces where residents come together to cultivate food and ornamental plants. These gardens foster community engagement, provide access to fresh produce, and create green spaces in urban environments.

Rooftop Farms

Rooftop farming utilizes the often underutilized space on building rooftops to grow vegetables, herbs, and fruits. These farms can be found on residential, commercial, and institutional buildings, contributing to local food production and urban greening.

Vertical Farming

Vertical farming involves growing crops in vertically stacked layers, often in controlled environments such as greenhouses or indoor facilities. This method maximizes space efficiency and allows for year-round production, making it ideal for densely populated urban areas.

Hydroponics and Aquaponics

Hydroponics is a soil-less method of growing plants using nutrient-rich water, while aquaponics combines hydroponics with aquaculture, raising fish alongside plants in a symbiotic system. Both methods are well-suited for urban environments, as they require less space and water than traditional agriculture.

Urban Farms and Greenhouses

Larger-scale urban farms and greenhouses can be found in and around cities, producing significant quantities of food for local markets. These farms often employ advanced agricultural techniques and technologies to optimize productivity and sustainability.

Benefits of Urban Agriculture

Enhanced Food Security

Local Food Production

Urban agriculture increases local food production, reducing reliance on external food sources and enhancing food security. By growing food within city limits, urban residents have greater access to fresh, nutritious produce.

Food Sovereignty

Urban agriculture empowers communities to take control of their food systems, promoting food sovereignty. This approach allows residents to produce culturally appropriate and diverse foods, preserving traditional diets and practices.

Environmental Sustainability

Reduced Food Miles

By producing food locally, urban agriculture reduces the distance food travels from farm to table, known as food miles. This reduction lowers greenhouse gas emissions associated with transportation and decreases the carbon footprint of food production.

Urban Greening

Urban farming contributes to urban greening, creating green spaces that improve air quality, reduce urban heat islands, and enhance biodiversity. These green spaces also provide recreational and aesthetic benefits to urban residents.

Waste Recycling

Urban agriculture can utilize organic waste from cities as compost, reducing landfill waste and improving soil fertility. Additionally, some urban farms employ waste-to-energy technologies, converting organic waste into biogas or other forms of renewable energy.

Social and Economic Benefits

Community Building

Community gardens and urban farms foster social interactions, strengthen community bonds, and promote a sense of ownership and pride among residents. These spaces often serve as venues for educational programs, cultural events, and social gatherings.

Job Creation and Economic Opportunities

Urban agriculture creates jobs and economic opportunities, from farming and food processing to marketing and distribution. It also supports local economies by providing fresh produce to local markets, restaurants, and consumers.

Health and Well-Being

Access to fresh, locally grown food improves dietary quality and public health. Urban agriculture also promotes physical activity, mental well-being, and stress reduction, contributing to overall community health.

Challenges and Barriers to Urban Agriculture

Space Constraints

Limited Land Availability

Urban areas often have limited land available for agriculture, competing with other land uses such as housing, industry, and infrastructure. Creative solutions, such as repurposing vacant lots and utilizing rooftops, are necessary to overcome this challenge.

Soil Contamination

Urban soils can be contaminated with heavy metals, pollutants, and other hazardous substances, posing risks to food safety and human health. Testing and remediating soils, or using raised beds and soil-less growing methods, are essential for safe urban farming.

Regulatory and Policy Issues

Zoning and Land Use Regulations

Zoning and land use regulations can restrict urban agriculture activities, limiting where and how food can be grown in cities. Advocating for supportive policies and zoning changes is crucial for the expansion of urban agriculture.

Food Safety Regulations

Ensuring food safety is a major concern for urban agriculture. Regulations governing the production, processing, and distribution of food must be adapted to accommodate urban farming practices while maintaining safety standards.

Economic and Financial Barriers

High Start-Up Costs

Establishing urban farms and gardens can require significant initial investments in infrastructure, equipment, and resources. Access to funding, grants, and financial support is vital for the viability of urban agriculture projects.

Market Access and Distribution

Urban farmers often face challenges in accessing markets and distributing their produce. Developing local food networks, farmers' markets, and community-supported agriculture (CSA) programs can help urban farmers reach consumers and sustain their operations.

Technical and Knowledge Barriers

Lack of Agricultural Knowledge

Urban residents may lack the agricultural knowledge and skills necessary for successful farming. Education and training programs, as well as access to expert advice and resources, are essential for building the capacity of urban farmers.

Technological Constraints

While advanced technologies can enhance urban agriculture, their high cost and complexity can be barriers to adoption. Simplifying technologies and making them more accessible to urban farmers is important for widespread implementation.

Case Studies of Urban Agriculture

Detroit, USA: Urban Farming for Revitalization

Detroit has become a model for urban agriculture as a means of revitalizing post-industrial cities. Community groups, non-profits, and entrepreneurs have transformed vacant lots into productive urban farms, creating jobs, improving food security, and fostering community engagement.

Havana, Cuba: Urban Agriculture for Food Security

Havana's urban agriculture movement emerged in response to severe food shortages during the 1990s. The city developed a network of urban farms and gardens, producing a significant portion of its food supply and promoting sustainable agricultural practices.

Singapore: High-Tech Urban Farming

Singapore, with its limited land and high population density, has embraced high-tech urban farming solutions such as vertical farming and rooftop greenhouses. These innovations enhance food security, reduce food imports, and promote environmental sustainability.

Nairobi, Kenya: Community Gardens for Empowerment

In Nairobi, community gardens have been established in informal settlements, providing residents with access to fresh produce and income-generating opportunities. These gardens also serve as platforms for education and community development.

Future Trends and Potential of Urban Agriculture

Integration with Smart Cities

Urban agriculture can be integrated into the broader concept of smart cities, utilizing digital technologies and data analytics to optimize food production and distribution. Smart urban farms can contribute to the sustainability and resilience of urban areas.

Expansion of Vertical Farming

Vertical farming is expected to expand, driven by advancements in technology and increasing demand for local food production. These farms can be integrated into urban infrastructure, such as skyscrapers and multi-use buildings, maximizing space efficiency.

Climate Resilience and Adaptation

Urban agriculture can play a crucial role in enhancing climate resilience and adaptation. By diversifying food sources, improving urban green infrastructure, and reducing reliance on long food supply chains, cities can become more resilient to climate change impacts.

Policy and Institutional Support

Increased policy and institutional support will be essential for the growth of urban agriculture. Governments and organizations must develop supportive policies, provide funding and resources, and facilitate knowledge sharing and collaboration.

Community Engagement and Education

Engaging communities and providing education on urban agriculture will be key to its success. Programs that teach urban residents how to grow food, understand food systems, and appreciate the benefits of urban farming can foster a culture of sustainability and resilience.

Conclusion

Urban agriculture offers a transformative approach to addressing some of the most pressing challenges faced by modern cities, including food security, environmental sustainability, and community well-being. By integrating food production into urban landscapes, cities can enhance their resilience, reduce their ecological footprint, and foster stronger, healthier communities. Despite challenges such as space constraints, regulatory issues, and economic barriers, the potential benefits of urban agriculture make it a vital component of sustainable urban development. With continued innovation, policy support, and community engagement, urban agriculture can help create more productive, sustainable, and resilient urban environments for future generations.

References

Hardman, M., Clark, A., & Sherriff, G. (2022). Mainstreaming urban agriculture: opportunities and barriers to upscaling city farming. Agronomy, 12(3), 601.

Kaufman, J. L., & Bailkey, M. (2000). Farming inside cities: Entrepreneurial urban agriculture in the United States (pp. 1-124). Cambridge, MA: Lincoln Institute of Land Policy.

Mougeot, L. J. (2006). Growing better cities: Urban agriculture for sustainable development. IDRC.

Orsini, F., Kahane, R., Nono-Womdim, R., & Gianquinto, G. (2013). Urban agriculture in the developing world: a review. Agronomy for sustainable development, 33, 695-720.

8

Organic Farming: Back to the Roots

Abstract

Organic farming represents a return to traditional agricultural practices that emphasize sustainability, environmental stewardship, and health. This article explores the principles and practices of organic farming, its historical context, and its benefits and challenges. By examining case studies and the global impact of organic agriculture, this comprehensive discussion highlights how organic farming contributes to soil health, biodiversity, and food security while addressing consumer demands for healthier food and environmental protection. It also considers future directions and innovations in organic farming, emphasizing its role in building a sustainable and resilient food system.

Introduction

Organic farming has garnered significant attention and support as a sustainable agricultural practice that prioritizes environmental health, animal welfare, and human health. It eschews synthetic chemicals and genetically modified organisms (GMOs) in favor of natural processes and materials. As the global community grapples with the consequences of industrial agriculture, including environmental degradation and health concerns, organic farming offers a viable alternative that aligns with ecological and social values. This article delves into the essence of organic farming, tracing its historical roots, examining its core principles, and assessing its impact on modern agriculture.

Historical Context

Early Agricultural Practices

Agriculture has always been inherently organic. Before the advent of synthetic fertilizers and pesticides in the 20th century, farmers relied on natural methods to cultivate crops and raise livestock. Traditional farming practices, such as crop rotation, composting, and the use of natural pest predators, were integral to maintaining soil fertility and controlling pests.

The Rise of Industrial Agriculture

The Green Revolution of the mid-20th century marked a significant shift towards industrial agriculture. The introduction of high-yield crop varieties, synthetic fertilizers, pesticides, and advanced irrigation techniques dramatically increased food production. However, this shift also led to environmental problems, including soil degradation, water pollution, and loss of biodiversity.

Emergence of the Organic Movement

In response to the negative impacts of industrial agriculture, the organic movement began to take shape in the early 20th century. Pioneers like Sir Albert Howard in the UK and J.I. Rodale in the US advocated for a return to natural farming practices. The organic farming movement gained momentum in the 1960s and 1970s, coinciding with the broader environmental movement and growing public awareness of health and environmental issues.

Principles and Practices of Organic Farming

Soil Health and Fertility

Organic Matter and Composting

Soil health is the cornerstone of organic farming. Organic farmers enhance soil fertility by adding organic matter, such as compost and green manure. These materials improve soil structure, water retention, and nutrient availability.

Crop Rotation and Diversity

Crop rotation involves growing different types of crops in succession on the same land. This practice helps prevent soil depletion, reduces pest and disease buildup, and enhances soil fertility. Organic farmers also cultivate a diverse range of crops to promote ecosystem balance and resilience.

Pest and Disease Management

Biological Control

Organic farming relies on biological control methods to manage pests and diseases. This includes introducing natural predators, such as ladybugs to control aphids, and using beneficial microorganisms to suppress soil-borne diseases.

Cultural Practices

Cultural practices, such as intercropping and using resistant crop varieties, are essential components of organic pest management. These practices create an environment that is less conducive to pests and diseases.

Animal Welfare

Natural Living Conditions

Organic farming emphasizes the humane treatment of livestock. Animals are provided with natural living conditions, including access to outdoor spaces, organic feed, and opportunities to engage in natural behaviors.

Disease Prevention

Preventing disease in organic livestock involves promoting strong immune systems through good nutrition, adequate housing, and stress reduction. Organic farmers avoid the routine use of antibiotics and growth hormones.

Environmental Sustainability

Biodiversity Conservation

Organic farming practices promote biodiversity both on and off the farm. By avoiding synthetic chemicals and fostering diverse crop rotations, organic farms support a variety of plant and animal species.

Water Management

Organic farmers implement water conservation techniques, such as mulching and drip irrigation, to optimize water use and reduce runoff. These practices help protect water quality and reduce the farm's overall water footprint.

Benefits of Organic Farming

Environmental Benefits

Soil Health

Organic farming enhances soil health by increasing organic matter content, improving soil structure, and fostering beneficial soil microorganisms. Healthy soils are more resilient to erosion and better at sequestering carbon, contributing to climate change mitigation.

Reduced Chemical Pollution

By avoiding synthetic pesticides and fertilizers, organic farming reduces chemical runoff into waterways and decreases the risk of soil and water contamination. This benefits both ecosystems and human communities.

Biodiversity

Organic farms typically have higher biodiversity than conventional farms. The use of diverse crop rotations and natural pest management practices supports a variety of plant and animal species, enhancing ecosystem health.

Health Benefits

Nutrient-Rich Food

Studies have shown that organic foods often contain higher levels of certain nutrients, such as antioxidants, vitamins, and minerals, compared to conventionally grown foods. This is attributed to the healthier soils and absence of synthetic chemicals.

Reduced Exposure to Pesticides

Organic farming eliminates the use of synthetic pesticides, reducing the risk of pesticide residues on food. This decreases the potential for negative health effects associated with pesticide exposure, particularly for vulnerable populations like children.

Economic and Social Benefits

Sustainable Livelihoods

Organic farming can provide sustainable livelihoods for farmers by reducing input costs and providing access to premium markets. Organic products often command higher prices, offering economic incentives for farmers to adopt organic practices.

Community Building

Organic farming fosters strong community connections through farmers' markets, community-supported agriculture (CSA) programs, and local food networks. These connections support local economies and promote social cohesion.

Challenges and Criticisms of Organic Farming

Yield Comparisons

Lower Yields

One of the main criticisms of organic farming is that it generally produces lower yields compared to conventional farming. This can be a significant drawback, especially in the context of a growing global population and the need to ensure food security.

Yield Variability

Organic yields can be more variable and susceptible to weather conditions, pests, and diseases. This variability poses a risk to farmers' incomes and can affect the stability of food supply chains.

Economic Barriers

Higher Costs

Organic farming often involves higher labor and management costs due to the need for manual weeding, diverse crop rotations, and other labor-intensive practices. These higher costs can be a barrier for small-scale and resource-limited farmers.

Market Access

Access to markets for organic products can be challenging, particularly in regions where organic certification and consumer demand are limited. Farmers may struggle to find reliable markets and achieve premium prices for their products.

Certification and Regulation

Complexity of Certification

Obtaining organic certification can be a complex and costly process, involving extensive documentation and compliance with strict standards. This can be a barrier for small-scale and beginning farmers.

Variability in Standards

Organic standards and regulations vary by country and region, leading to inconsistencies and confusion in the marketplace. This variability can affect the credibility and consumer trust in organic products.

Global Impact of Organic Farming

Organic Agriculture in Europe

Policy Support

Europe has been a leader in promoting organic agriculture through policy support, subsidies, and research funding. The European Union's Common Agricultural Policy (CAP) includes provisions to support organic farming, contributing to its growth across the continent.

Market Demand

Europe has a strong market demand for organic products, driven by consumer preferences for healthier and more environmentally friendly foods. This demand has created robust markets for organic farmers and processors.

Organic Agriculture in North America

Market Growth

The organic food market in North America, particularly in the United States and Canada, has seen significant growth over the past few decades. Consumers are increasingly seeking out organic products, driving expansion in organic acreage and production.

Research and Innovation

North America is home to numerous research institutions and organizations dedicated to advancing organic agriculture. Innovations in organic farming techniques and technologies continue to emerge from this region.

Organic Agriculture in Developing Countries

Opportunities for Smallholders

Organic farming offers opportunities for smallholder farmers in developing countries to improve their livelihoods and access premium markets. Organic practices can be well-suited to resource-limited environments where synthetic inputs are unaffordable or unavailable.

Challenges and Support

Developing countries face challenges in expanding organic agriculture, including limited access to knowledge, resources, and certification services. International development programs and NGOs play a crucial role in supporting organic farming initiatives.

Case Studies

The Rodale Institute, USA

Pioneering Organic Research

The Rodale Institute has been at the forefront of organic research and advocacy since its founding in 1947. Its Farming Systems Trial, the longest-running comparison of organic and conventional farming systems in North America, provides valuable data on the benefits and challenges of organic agriculture.

Education and Outreach

The Rodale Institute offers education and training programs for farmers, researchers, and consumers, promoting the adoption of organic practices and raising awareness about the benefits of organic farming.

Sekem Initiative, Egypt

Holistic Sustainable Development

The Sekem Initiative in Egypt integrates organic farming with social, cultural, and economic development. Founded in 1977, Sekem has transformed desert land into productive organic farms, while also establishing schools, medical centers, and cultural institutions.

Biodynamic Farming

Sekem practices biodynamic farming, a holistic approach that views the farm as a self-sustaining ecosystem. Biodynamic methods include the use of natural preparations, composting, and crop rotations to enhance soil health and biodiversity.

Navdanya, India

Seed Sovereignty and Agroecology

Navdanya, founded by Dr. Vandana Shiva in 1987, promotes seed sovereignty and agroecology in India. The organization supports farmers in adopting organic practices, conserving indigenous seed varieties, and resisting the spread of GMOs.

Training and Advocacy

Navdanya provides training for farmers on organic methods and advocates for policies that support sustainable agriculture. Its work has empowered thousands of farmers to transition to organic farming and protect their agricultural heritage.

Future Directions and Innovations in Organic Farming

Technological Advancements

Precision Agriculture

Precision agriculture technologies, such as GPS-guided equipment and drones, can enhance the efficiency and effectiveness of organic farming. These tools enable precise application of organic inputs, optimize irrigation, and monitor crop health.

Biopesticides and Biofertilizers

Advancements in biopesticides and biofertilizers offer organic farmers new tools for pest and nutrient management. These products, derived from natural sources, provide environmentally friendly alternatives to synthetic chemicals.

Policy and Institutional Support

Government Incentives

Government incentives, such as subsidies, grants, and tax breaks, can encourage the adoption of organic farming practices. Policies that support organic certification, research, and market development are crucial for the growth of organic agriculture.

International Collaboration

International collaboration and knowledge sharing can accelerate the spread of organic farming practices. Organizations such as the International Federation of Organic Agriculture Movements (IFOAM) facilitate global cooperation and advocacy for organic agriculture.

Consumer Education and Awareness

Promoting Organic Benefits

Educating consumers about the benefits of organic farming and organic products is essential for sustaining demand. Public awareness campaigns, labeling initiatives, and transparency in food production can build consumer trust and support for organic agriculture.

Addressing Misconceptions

Addressing misconceptions about organic farming, such as the belief that organic foods are inherently safer or more nutritious, is important for informed consumer choices. Providing accurate information about the benefits and limitations of organic farming can foster a balanced understanding.

Integrating Organic and Conventional Practices

Sustainable Intensification

Integrating organic and conventional practices through sustainable intensification can help meet global food needs while minimizing environmental impacts. This approach involves combining the best practices of both systems to enhance productivity and sustainability.

Agroecology

Agroecology, which emphasizes the ecological principles of agriculture, offers a framework for integrating organic and conventional practices. Agroecological approaches prioritize biodiversity, soil health, and ecosystem services, aligning with the goals of organic farming.

Conclusion

Organic farming represents a return to agricultural practices that respect natural processes and promote sustainability. By focusing on soil health, biodiversity, and the well-being of humans and animals, organic farming addresses many of the environmental and social challenges posed by industrial agriculture. Despite its challenges, organic farming offers significant benefits for the environment, human health, and rural livelihoods. As consumer demand for organic products continues to grow, and as innovations and policies support the expansion of organic agriculture, organic farming is poised to play a crucial role in building a sustainable and resilient global food system. Through continued research, education, and collaboration, the organic movement can help shape a future where agriculture works in harmony with nature.

References

Barton, G. A. (2018). The global history of organic farming. Oxford University Press.

Heckman, J. (2006). A history of organic farming: Transitions from Sir Albert Howard's War in the Soil to USDA National Organic Program. Renewable Agriculture and Food Systems, 21(3), 143-150.

Ramanathan, K. M. (2006). Organic farming for sustainability. Journal of the Indian Society of Soil Science, 54(4), 418-425.

Siddique, S., Hamid, M., Tariq, A., & Kazi, A. G. (2014). Organic farming: the return to nature. Improvement of Crops in the Era of Climatic Changes: Volume 2, 249-281.

Vogt, G. (2007). The origins of organic farming. Organic farming: An international history, 1, 9-29.

9

The Global Food Supply Chain From Farm to Table

Abstract

The global food supply chain is a complex system that encompasses all stages of food production, processing, distribution, and consumption. This article provides an in-depth exploration of the journey food takes from farms to tables worldwide. It examines the various components of the supply chain, including agricultural production, food processing, logistics, retail, and consumption. The article also addresses key challenges such as food security, sustainability, and safety, and discusses the impact of globalization and technological advancements. Through case studies and examples, the article highlights the interconnected nature of the food supply chain and emphasizes the importance of cooperation among stakeholders to ensure a resilient and sustainable global food system.

Introduction

The global food supply chain is essential for ensuring that the growing population has access to sufficient, safe, and nutritious food. This intricate network involves multiple stages, from agricultural production to food processing, distribution, retail, and finally, consumption. Understanding how these stages interact and the challenges they face is crucial for developing strategies to improve food security, sustainability, and resilience. This article explores each component of the food supply chain, highlighting the processes, technologies, and issues that influence the journey of food from farm to table.

Agricultural Production

Crop Production

Traditional Farming Methods

Traditional farming methods have been the foundation of agricultural production for centuries. These methods include crop rotation, intercropping, and the use of organic fertilizers. They promote soil health, biodiversity, and sustainable land use.

Modern Agricultural Practices

Modern agriculture has seen the introduction of advanced techniques such as precision farming, genetically modified organisms (GMOs), and the use of synthetic fertilizers and pesticides. These methods aim to increase yield, efficiency, and resilience to climate change.

Sustainable Farming

Sustainable farming practices, such as organic farming, agroecology, and regenerative agriculture, focus on long-term environmental health. They aim to minimize negative impacts on the ecosystem while maintaining productivity and profitability.

Livestock Production

Conventional Livestock Farming

Conventional livestock farming involves raising animals in large numbers, often in confined spaces, with the use of growth hormones, antibiotics, and concentrated feed. This method prioritizes high production efficiency but raises concerns about animal welfare and environmental impacts.

Sustainable Livestock Practices

Sustainable livestock practices emphasize animal welfare, environmental stewardship, and social responsibility. Methods include free-range grazing, rotational grazing, and integrating livestock with crop production to create a balanced farm ecosystem.

Food Processing and Manufacturing

Primary Processing

Post-Harvest Handling

Post-harvest handling involves the initial steps of cleaning, sorting, and packing raw agricultural products. Proper handling is crucial to minimize losses and maintain quality.

Preservation Techniques

Various preservation techniques, such as drying, freezing, canning, and fermenting, are used to extend the shelf life of food products. These methods help reduce food waste and ensure a steady supply of food throughout the year.

Secondary Processing

Value-Added Products

Secondary processing transforms raw agricultural products into value-added food items, such as bread from wheat, cheese from milk, and canned vegetables. This stage increases the economic value of the products and diversifies the food supply.

Food Safety and Quality Control

Ensuring food safety and quality is a critical aspect of food processing. Regulations and standards, such as Hazard Analysis and Critical Control Points (HACCP) and Good Manufacturing Practices (GMP), are implemented to prevent contamination and ensure consumer safety.

Distribution and Logistics

Transportation

Modes of Transport

Food products are transported using various modes, including trucks, trains, ships, and airplanes. The choice of transport mode depends on factors such as distance, cost, and the nature of the food products.

Cold Chain Logistics

Cold chain logistics involves maintaining a temperature-controlled supply chain to preserve the quality and safety of perishable food products. This system includes refrigerated storage, transportation, and handling.

Warehousing

Storage Facilities

Warehousing plays a crucial role in the food supply chain by providing storage for raw materials, semi-processed, and finished food products. Proper storage conditions, such as temperature and humidity control, are essential to prevent spoilage and maintain quality.

Inventory Management

Effective inventory management ensures that the right quantity of food products is available at the right time and place. Techniques such as Just-In-Time (JIT) inventory and automated inventory systems help optimize stock levels and reduce waste.

Retail and Consumer Access

Supermarkets and Grocery Stores

Retail Chains

Supermarkets and grocery stores are the primary points of sale for food products. Large retail chains dominate the market, offering a wide variety of products and convenience to consumers.

Local Markets

Local markets provide an alternative to large retail chains, offering fresh, locally-produced food products. These markets support local economies and promote shorter supply chains.

Online Food Retail

E-Commerce Platforms

The rise of e-commerce platforms has transformed the food retail sector, providing consumers with the convenience of ordering food products online. These platforms offer a wide range of products and home delivery services.

Direct-to-Consumer Models

Direct-to-consumer models, such as subscription boxes and farm-to-table services, connect consumers directly with producers. This approach reduces intermediaries, supports local farmers, and ensures fresher products.

Consumption and Food Waste

Consumer Behavior

Food Preferences

Consumer food preferences are influenced by factors such as culture, income, health considerations, and environmental awareness. Understanding these preferences helps shape food production and marketing strategies.

Food Waste

Food waste is a significant issue in the global food supply chain. Consumers often discard food due to over-purchasing, poor storage, and misunderstanding of expiration dates. Reducing food waste requires awareness and behavioral changes at the consumer level.

Food Security

Accessibility and Affordability

Food security depends on the accessibility and affordability of nutritious food for all individuals. Efforts to improve food security include social safety nets, food assistance programs, and policies that support equitable food distribution.

Nutrition and Health

Ensuring that the food supply chain provides nutritious and safe food is essential for public health. Initiatives to improve nutrition include fortification, education programs, and promoting diverse and balanced diets.

Challenges in the Global Food Supply Chain

Climate Change

Impact on Agriculture

Climate change poses significant challenges to agriculture, including changing weather patterns, increased frequency of extreme events, and shifting growing seasons. These changes affect crop yields, livestock health, and food production stability.

Adaptation Strategies

Adaptation strategies, such as developing climate-resilient crop varieties, improving water management, and implementing sustainable farming practices, are essential to mitigate the impacts of climate change on the food supply chain.

Sustainability

Environmental Impact

The food supply chain has a considerable environmental impact, including greenhouse gas emissions, deforestation, water usage, and pollution. Sustainable practices aim to reduce these impacts and promote environmental health.

Sustainable Practices

Implementing sustainable practices, such as organic farming, agroforestry, and integrated pest management, can help create a more environmentally friendly food supply chain. These practices prioritize long-term ecological balance and resource conservation.

Food Safety

Contamination Risks

Food contamination risks arise at various stages of the supply chain, from production to consumption. Contaminants can include pathogens, chemicals, and physical hazards, posing health risks to consumers.

Regulatory Frameworks

Robust regulatory frameworks, such as the Food Safety Modernization Act (FSMA) in the US and the European Food Safety Authority (EFSA) in the EU, are crucial for ensuring food safety. These regulations establish standards and protocols to prevent and respond to contamination incidents.

Economic and Political Factors

Trade Policies

Trade policies influence the global food supply chain by affecting the flow of goods, tariffs, and market access. Trade agreements and disputes can have significant impacts on food prices and availability.

Economic Inequality

Economic inequality affects access to food and the ability to participate in the food supply chain. Addressing inequality involves promoting fair wages, supporting smallholder farmers, and ensuring inclusive economic growth.

The Impact of Globalization and Technology

Globalization

International Trade

Globalization has expanded international trade, enabling the movement of food products across borders. This has increased food availability and diversity but also introduced vulnerabilities related to supply chain disruptions.

Cultural Exchange

Globalization has facilitated cultural exchange, influencing food preferences and dietary patterns. This exchange has led to greater diversity in food offerings but also the homogenization of diets and the loss of traditional food practices.

Technological Advancements

Precision Agriculture

Precision agriculture uses technologies such as GPS, drones, and data analytics to optimize farming practices. These technologies improve efficiency, reduce resource use, and enhance crop management.

Blockchain and Traceability

Blockchain technology enhances traceability in the food supply chain by providing a secure and transparent record of transactions. This technology improves food safety, reduces fraud, and builds consumer trust.

Case Studies

The Netherlands: Sustainable Agriculture and Innovation

Greenhouse Horticulture

The Netherlands is a global leader in greenhouse horticulture, producing high yields of vegetables, flowers, and plants in controlled environments. This approach maximizes resource efficiency and minimizes environmental impact.

Circular Agriculture

The Dutch concept of circular agriculture focuses on closing nutrient loops, reducing waste, and integrating crop and livestock production. This holistic approach promotes sustainability and resilience.

Brazil: Soybean Production and Deforestation

Agricultural Expansion

Brazil is one of the world's largest producers of soybeans, a key commodity in the global food supply chain. However, agricultural expansion has led to deforestation in the Amazon rainforest, raising environmental concerns.

Sustainable Soy Initiatives

Efforts to promote sustainable soy production include certification schemes, such as the Round Table on Responsible Soy (RTRS), and government policies to protect forests and promote sustainable land use.

India: Smallholder Farmers and Food Security

Diverse Agricultural Systems

India's agricultural landscape is characterized by diverse farming systems, with millions of smallholder farmers growing a wide variety of crops. These systems contribute to food security and cultural heritage.

Challenges and Solutions

Smallholder farmers face challenges such as limited access to resources, climate change, and market volatility. Solutions include improving access to technology, providing financial support, and strengthening farmer cooperatives.

Future Directions and Innovations

Resilient Food Systems

Diversification

Diversifying crops, production systems, and supply chains enhances resilience by reducing dependence on a single source or method. Diversification strategies include intercropping, polyculture, and supporting local food systems.

Adaptive Capacity

Building adaptive capacity involves developing the ability to respond to changes and shocks in the food supply chain. This includes investing in research, education, and infrastructure to support adaptation and innovation.

Sustainable Development Goals (SDGs)

Zero Hunger

Achieving the SDG of zero hunger requires addressing food security, nutrition, and sustainable agriculture. Efforts include promoting equitable access to food, reducing food waste, and supporting sustainable farming practices.

Climate Action

Integrating climate action into the food supply chain involves reducing greenhouse gas emissions, enhancing carbon sequestration, and promoting climate-resilient practices. This contributes to broader efforts to mitigate and adapt to climate change.

Policy and Governance

International Cooperation

International cooperation is essential for addressing global food supply chain challenges. Multilateral organizations, such as the Food and Agriculture Organization (FAO) and the World Trade Organization (WTO), play key roles in facilitating collaboration and setting standards.

National and Local Policies

National and local policies that support sustainable agriculture, food security, and economic equity are crucial for a resilient food supply chain. These policies include subsidies, research funding, and regulations that promote sustainable practices.

Conclusion

The global food supply chain is a complex and dynamic system that plays a critical role in feeding the world's population. Understanding the intricacies of this system, from agricultural production to consumption, is essential for addressing the challenges and opportunities it presents. By embracing sustainable practices, leveraging technological advancements, and fostering international cooperation, stakeholders can work together to create a resilient and sustainable global food supply chain. Ensuring that this system is robust and adaptable is vital for meeting the needs of current and future generations, promoting food security, and protecting the environment.

References

Bhat, R. V. (2013). Achieving food safety from farm to table: global requirements and the Indian scenario. Bulletin of the Nutrition Foundation of India, 34(4).

Dani, S. (2015). Food supply chain management and logistics: From farm to fork. Kogan Page Publishers.

Donaldson, A. (2022). Digital from farm to fork: Infrastructures of quality and control in food supply chains. Journal of Rural Studies, 91, 228-235.

Gharehgozli, A., Iakovou, E., Chang, Y., & Swaney, R. (2017). Trends in global E-food supply chain and implications for transport: Literature review and research directions. Research in transportation business & management, 25, 2-14.

Moysiadis, T., Spanaki, K., Kassahun, A., Kläser, S., Becker, N., Alexiou, G., ... & Karali, I. (2023). AgriFood supply chain traceability: data sharing in a farm-to-fork case. Benchmarking: An International Journal, 30(9), 3090-3123.

Yi, J., Meemken, E. M., Mazariegos-Anastassiou, V., Liu, J., Kim, E., Gómez, M. I., ... & Barrett, C. B. (2021). Post-farmgate food value chains make up most of consumer food expenditures globally. Nature Food, 2(6), 417-425.

10

Agricultural Policies and Their Implications

Abstract

Agricultural policies shape the landscape of farming and food production globally. They encompass a range of interventions by governments to influence agricultural practices, food security, trade, and rural development. This article delves into the various types of agricultural policies, their historical context, and their socio-economic and environmental implications. It examines case studies from different regions to highlight the diverse approaches and outcomes of these policies. Additionally, the article discusses the challenges and opportunities in formulating and implementing effective agricultural policies, particularly in the face of global issues such as climate change, food insecurity, and market volatility. By understanding the impact of agricultural policies, stakeholders can better navigate the complexities of the agricultural sector and promote sustainable development.

Introduction

Agricultural policies are a critical component of national and international strategies for food production, rural development, and economic growth. They encompass a wide range of measures, including subsidies, tariffs, price controls, research funding, and land use regulations. These policies aim to enhance productivity, ensure food security, stabilize markets, and promote sustainable agricultural practices. However, the effectiveness and impact of these policies vary widely depending on the context and implementation. This article explores the different dimensions of agricultural policies, their historical evolution, and their implications for various stakeholders.

Historical Context of Agricultural Policies

Early Agricultural Policies

Agricultural policies have been a part of human society since the dawn of civilization. Early policies focused on land distribution, irrigation, and storage to ensure food availability and stability. Ancient civilizations, such as those in Mesopotamia, Egypt, and China, developed sophisticated agricultural systems supported by state policies.

Industrial Revolution and Agricultural Policy

The Industrial Revolution brought significant changes to agriculture, with policies shifting towards modernization and industrialization. Governments in Europe and North America introduced measures to increase agricultural productivity through mechanization, improved crop varieties, and chemical fertilizers.

Post-World War II Agricultural Policies

After World War II, agricultural policies in many developed countries focused on food security and rural development. The United States, through its Farm Bill, and the European Union, with the Common Agricultural Policy (CAP), implemented comprehensive policies to support farmers, stabilize prices, and ensure food supplies.

Types of Agricultural Policies

Subsidies and Support Payments

Direct Payments

Direct payments to farmers are one of the most common forms of agricultural subsidies. These payments aim to support farmers' incomes, stabilize agricultural production, and promote rural development. They can be based on factors such as land area, crop type, or historical production levels.

Price Supports

Price supports involve government intervention to maintain prices of agricultural products at a certain level. This can include purchasing surplus production, setting minimum prices, or providing compensation when market prices fall below a certain threshold.

Trade Policies

Tariffs and Quotas

Tariffs and quotas are used to protect domestic agriculture from foreign competition. Tariffs impose taxes on imported goods, while quotas limit the quantity of certain imports. These measures aim to support domestic farmers and ensure food security.

Export Subsidies

Export subsidies are financial incentives provided to domestic producers to encourage the export of agricultural products. These subsidies aim to enhance the competitiveness of domestic products in international markets.

Research and Development

Agricultural Research Funding

Government funding for agricultural research is essential for developing new technologies, improving crop varieties, and enhancing sustainable practices. Research institutions, universities, and extension services play key roles in advancing agricultural knowledge.

Innovation and Technology Transfer

Policies that promote innovation and technology transfer help farmers adopt new practices and technologies. This includes funding for research, development of extension services, and partnerships between public and private sectors.

Environmental and Sustainability Policies

Conservation Programs

Conservation programs aim to protect natural resources, such as soil, water, and biodiversity. These programs can include incentives for adopting sustainable practices, such as crop rotation, cover cropping, and reduced tillage.

Climate Change Mitigation

Policies that address climate change in agriculture focus on reducing greenhouse gas emissions, enhancing carbon sequestration, and improving resilience to climate impacts. This includes support for renewable energy, sustainable land management, and climate-smart agriculture practices.

Rural Development Policies

Infrastructure Development

Investing in rural infrastructure, such as roads, irrigation systems, and storage facilities, is crucial for improving agricultural productivity and market access. These investments help reduce post-harvest losses and enhance food security.

Social Services

Providing social services, such as education, healthcare, and financial services, in rural areas is essential for improving the quality of life for farming communities. These services support rural development and reduce urban-rural disparities.

Socio-Economic Implications of Agricultural Policies

Income and Livelihoods

Farmer Incomes

Agricultural policies that provide direct payments, price supports, and other subsidies help stabilize and increase farmers' incomes. This financial support is crucial for smallholder farmers who are vulnerable to market fluctuations and climate impacts.

Rural Employment

Policies that promote agricultural development and rural infrastructure create employment opportunities in rural areas. This includes jobs in farming, processing, transportation, and retail, contributing to rural economic growth.

Food Security

Availability

Agricultural policies play a vital role in ensuring the availability of food by supporting domestic production and stabilizing markets. Policies that promote research and innovation also contribute to increased productivity and food supply.

Accessibility

Policies that support rural development, improve infrastructure, and provide social services enhance food accessibility for rural communities. Trade policies that ensure fair and open markets also contribute to global food accessibility.

Market Stability

Price Volatility

Agricultural policies that include price supports, strategic reserves, and market interventions help stabilize prices and reduce volatility. This stability is crucial for farmers, consumers, and the overall economy.

Supply Chain Resilience

Policies that enhance supply chain resilience, such as investments in infrastructure and support for diversified farming systems, reduce the risk of disruptions. This resilience is essential for maintaining food security and market stability.

Environmental Implications of Agricultural Policies

Land Use and Management

Sustainable Land Use

Policies that promote sustainable land use practices, such as conservation tillage, agroforestry, and integrated pest management, help protect natural resources and maintain ecosystem services.

Deforestation and Land Degradation

Policies that fail to address unsustainable practices can lead to deforestation, land degradation, and loss of biodiversity. Effective policies must balance agricultural production with environmental conservation.

Water Management

Irrigation and Water Use Efficiency

Policies that support efficient water use, such as drip irrigation, rainwater harvesting, and watershed management, are crucial for sustainable agriculture. These measures help conserve water resources and ensure long-term productivity.

Pollution and Water Quality

Agricultural policies must address the impact of farming practices on water quality. This includes regulating the use of fertilizers and pesticides, promoting organic farming, and implementing buffer zones to protect water bodies.

Climate Change Mitigation and Adaptation

Greenhouse Gas Emissions

Policies that promote low-emission farming practices, such as reduced tillage, cover cropping, and agroforestry, help mitigate greenhouse gas emissions from agriculture. Support for renewable energy and carbon sequestration initiatives also contributes to climate change mitigation.

Adaptation Strategies

Policies that enhance farmers' resilience to climate change, such as crop insurance, climate-smart agriculture, and disaster risk management, are essential for ensuring food security in a changing climate.

Case Studies

The European Union's Common Agricultural Policy (CAP)

Overview

The Common Agricultural Policy (CAP) is one of the most comprehensive agricultural policy frameworks in the world. It aims to support farmers, ensure food security, and promote sustainable agriculture across EU member states.

Impact

CAP has significantly influenced agricultural practices in Europe, contributing to increased productivity and rural development. However, it has also faced criticism for promoting intensive farming practices and unequal distribution of subsidies.

The United States Farm Bill

Overview

The Farm Bill is a multi-year legislation that governs agricultural and food programs in the United States. It includes provisions for farm subsidies, conservation programs, research funding, and nutrition assistance.

Impact

The Farm Bill has played a crucial role in stabilizing farm incomes, promoting conservation practices, and supporting rural communities. However, it has also been criticized for favoring large-scale farms and not adequately addressing environmental concerns.

India's National Agricultural Policy

Overview

India's National Agricultural Policy aims to achieve sustainable agriculture, enhance food security, and improve farmers' livelihoods. Key components include support for research and extension, price supports, and rural development initiatives.

Impact

India's agricultural policies have contributed to significant improvements in food production and rural development. However, challenges remain in addressing smallholder farmers' needs, market access, and climate resilience.

Challenges in Agricultural Policy Formulation and Implementation

Policy Coherence

Intersectoral Coordination

Effective agricultural policies require coordination across sectors, such as water, energy, and environment. Ensuring policy coherence and avoiding contradictory measures is crucial for achieving sustainable outcomes.

Multi-Level Governance

Agricultural policies must be implemented at multiple levels, from local to global. Ensuring alignment and cooperation among different governance levels is essential for effective policy implementation.

Equity and Inclusiveness

Smallholder Farmers

Policies must address the needs of smallholder farmers, who are often marginalized and vulnerable to market and climate risks. This includes providing access to resources, markets, and support services.

Gender and Social Inclusion

Agricultural policies should promote gender equality and social inclusion by addressing the specific needs and challenges faced by women and marginalized groups in agriculture.

Monitoring and Evaluation

Impact Assessment

Regular monitoring and evaluation of agricultural policies are essential for assessing their impact and effectiveness. This includes collecting data, conducting research, and engaging stakeholders in the evaluation process.

Adaptive Management

Policies must be flexible and adaptive to changing conditions and new information. Adaptive management involves continuously learning from policy implementation and making necessary adjustments.

Opportunities for Improving Agricultural Policies

Integrating Sustainability

Agroecology and Sustainable Intensification

Promoting agroecology and sustainable intensification practices can enhance productivity while maintaining environmental health. Policies should support research, education, and incentives for adopting these practices.

Ecosystem-Based Approaches

Ecosystem-based approaches, such as landscape restoration and watershed management, integrate agricultural production with conservation goals. Policies should promote these holistic approaches to ensure long-term sustainability.

Leveraging Technology and Innovation

Digital Agriculture

Digital technologies, such as precision farming, remote sensing, and data analytics, offer opportunities for improving agricultural productivity and sustainability. Policies should support the development and adoption of these technologies.

Biotechnology and Genetic Engineering

Biotechnology and genetic engineering can contribute to improved crop varieties, pest resistance, and climate resilience. Policies should ensure responsible use, regulatory oversight, and public acceptance of these technologies.

Strengthening International Cooperation

Global Policy Frameworks

International frameworks, such as the Sustainable Development Goals (SDGs) and the Paris Agreement, provide a basis for coordinated action on agricultural policy. Countries should align their policies with these global commitments.

Trade and Market Access

Promoting fair and open trade can enhance global food security and market stability. Policies should support international cooperation, reduce trade barriers, and ensure equitable market access for all countries.

Conclusion

Agricultural policies are vital for shaping the future of food production, rural development, and environmental sustainability. By understanding the historical context, types, and implications of these policies, stakeholders can work towards creating a more resilient and equitable agricultural system. Addressing the challenges in policy formulation and implementation, while leveraging opportunities for innovation and sustainability, is essential for achieving global food security and sustainable development. Through collaboration and informed decision-making, agricultural policies can contribute to a better future for farmers, consumers, and the planet.

References

Bergmann, D. (1977). Agricultural policies in the EEC and their external implications. World development, 5(5-7), 407-415.

Bonnen, J. T. (1973). Implications for agricultural policy. American journal of agricultural economics, 55(3), 391-398.

Ellis, F. (1992). Agricultural policies in developing countries. Cambridge university press.

Hall, A., Mytelka, L., & Oyeyinka, B. (2005). Innovation systems: Implications for agricultural policy and practice.

Janvry, A. D. (1985). Integration of agriculture in the national and world economy: Implications for agricultural policies in developing countries.

11

The Economics of Modern Farming

Abstract

The economics of modern farming is a dynamic and complex field, reflecting the intricate interplay between technological advancements, market forces, policy frameworks, and environmental considerations. This article explores the multifaceted economic dimensions of contemporary agriculture, including production costs, market structures, supply chain dynamics, and the impact of government policies. It examines how modern farming practices, such as precision agriculture and biotechnology, influence productivity and profitability. The article also delves into the challenges and opportunities faced by farmers in today's globalized economy, emphasizing the need for sustainable practices and resilience to climate change. By understanding the economic principles underlying modern farming, stakeholders can better navigate the evolving agricultural landscape and promote a more sustainable and profitable farming sector.

Introduction

The economic landscape of modern farming has undergone significant transformations over the past few decades. Advances in technology, globalization, changing consumer preferences, and environmental challenges have reshaped agricultural practices and market dynamics. Understanding the economics of modern farming involves examining the costs and revenues associated with agricultural production, the structure of agricultural markets, and the impact of policies and external factors on farming operations. This article provides an in-depth analysis of these aspects, highlighting the economic principles and trends that influence modern farming.

The Evolution of Modern Farming

Technological Advancements

Precision Agriculture

Precision agriculture uses technologies such as GPS, drones, and data analytics to enhance farm management. These tools enable farmers to monitor crop conditions, optimize resource use, and improve yields. The economic benefits include reduced input costs, increased efficiency, and higher productivity.

Biotechnology

Biotechnology involves the use of genetic engineering, molecular markers, and other biotechnological tools to develop improved crop varieties and livestock breeds. These advancements can lead to higher yields, pest and disease resistance, and improved nutritional quality, thereby increasing profitability.

Market Dynamics

Globalization

Globalization has expanded markets for agricultural products, providing farmers with opportunities to sell their produce internationally. However, it also exposes them to global market fluctuations, trade policies, and competition from foreign producers.

Consumer Preferences

Changing consumer preferences, such as the demand for organic and locally-sourced food, have influenced agricultural practices and market opportunities. Farmers must adapt to these trends to remain competitive and meet consumer expectations.

Policy Frameworks

Agricultural Subsidies

Government subsidies play a significant role in supporting farmers' incomes, stabilizing prices, and encouraging sustainable practices. However, the distribution and impact of subsidies can vary, influencing the economic viability of different types of farming operations.

Trade Policies

Trade policies, including tariffs, quotas, and trade agreements, affect the competitiveness of agricultural products in global markets. These policies can either support or hinder farmers' access to international markets, impacting their economic outcomes.

Production Economics

Cost Structures

Variable Costs

Variable costs in farming include expenses that fluctuate with the level of production, such as seeds, fertilizers, pesticides, water, and labor. Managing these costs efficiently is crucial for maintaining profitability.

Fixed Costs

Fixed costs are expenses that remain constant regardless of the level of production, such as land, equipment, and infrastructure. Understanding and optimizing fixed costs is essential for long-term economic sustainability.

Revenue Streams

Crop Sales

The primary revenue stream for crop farmers is the sale of harvested crops. Prices can vary based on market demand, quality, and timing of sales. Diversifying crops can help mitigate risks associated with price volatility.

Livestock Sales

For livestock farmers, revenue comes from the sale of animals and animal products, such as meat, milk, and eggs. Market prices for livestock can be influenced by factors such as feed costs, disease outbreaks, and consumer demand.

Value-Added Products

Value-added products, such as processed foods, organic produce, and specialty items, can generate higher revenues compared to raw agricultural products. Farmers can enhance profitability by investing in processing and marketing these products.

Market Structures and Supply Chains

Agricultural Markets

Commodity Markets

Commodity markets for agricultural products are characterized by standardized products and significant price volatility. Prices are often determined by supply and demand dynamics, influenced by factors such as weather, global production levels, and trade policies.

Niche Markets

Niche markets, such as organic, fair trade, and local food markets, cater to specific consumer preferences and can offer higher prices. Farmers targeting niche markets must meet stringent standards and invest in marketing to differentiate their products.

Supply Chain Dynamics

Input Suppliers

Input suppliers provide essential products such as seeds, fertilizers, pesticides, machinery, and technology. The relationships between farmers and input suppliers can impact production costs and access to innovations.

Distribution Channels

Distribution channels include wholesalers, retailers, and direct-to-consumer sales. The choice of distribution channel affects the prices farmers receive and their ability to reach consumers. Direct sales, such as farmers' markets and CSA (Community Supported Agriculture) programs, can provide higher margins but require more marketing effort.

Retailers and Consumers

Retailers play a crucial role in connecting farmers with consumers. Large retailers can influence prices and demand for agricultural products, while consumer preferences drive trends in food production and marketing.

Government Policies and Their Impact

Agricultural Subsidies

Types of Subsidies

Government subsidies can take various forms, including direct payments, crop insurance, price supports, and conservation incentives. These subsidies aim to stabilize farm incomes, encourage sustainable practices, and ensure food security.

Impact on Farming Practices

Subsidies can influence farmers' decisions regarding crop selection, land use, and investment in technology. While subsidies can provide financial stability, they can also lead to overproduction and environmental degradation if not managed properly.

Trade Policies

Tariffs and Quotas

Tariffs and quotas are used to protect domestic agriculture from foreign competition. These measures can support local farmers by providing price stability and market access but can also lead to trade disputes and market distortions.

Trade Agreements

Trade agreements, such as NAFTA and the EU's Common Agricultural Policy, facilitate the exchange of agricultural products between countries. These agreements can open new markets for farmers but also increase competition and exposure to global market fluctuations.

Environmental Regulations

Sustainability Standards

Environmental regulations set standards for sustainable farming practices, such as soil conservation, water management, and pesticide use. Compliance with these standards can increase production costs but also promote long-term sustainability and access to premium markets.

Climate Change Policies

Policies addressing climate change, such as carbon taxes and incentives for renewable energy, impact farming operations. Farmers must adapt to these policies by adopting practices that reduce emissions and enhance resilience to climate impacts.

Challenges and Opportunities

Economic Challenges

Market Volatility

Price volatility in agricultural markets poses significant risks to farmers' incomes. Factors such as weather, global supply and demand, and trade policies contribute to unpredictable market conditions.

Access to Capital

Access to capital is crucial for investing in modern farming technologies and practices. Small and medium-sized farms often face challenges in securing financing due to high costs and perceived risks.

Environmental Challenges

Climate Change

Climate change impacts agricultural production through changing weather patterns, extreme events, and shifting growing seasons. Farmers must adapt to these changes to maintain productivity and profitability.

Resource Depletion

The depletion of natural resources, such as soil fertility and water availability, threatens the sustainability of modern farming. Sustainable practices and efficient resource management are essential to address these challenges.

Technological Opportunities

Innovation and Efficiency

Technological innovations, such as automation, artificial intelligence, and biotechnology, offer opportunities to enhance productivity, reduce costs, and improve sustainability. Investing in these technologies can lead to significant economic benefits.

Data-Driven Decision Making

The use of big data and analytics enables farmers to make informed decisions based on real-time information. This approach can optimize resource use, improve yields, and enhance profitability.

Market Opportunities

Diversification

Diversifying crops and revenue streams can help farmers mitigate risks associated with market volatility and climate change. Opportunities include value-added products, agritourism, and renewable energy production.

Sustainable and Ethical Markets

Growing consumer demand for sustainable and ethically-produced food presents opportunities for farmers to access premium markets. Adopting sustainable practices and obtaining certifications can enhance marketability and profitability.

Case Studies

The United States: Precision Agriculture

Adoption and Impact

Precision agriculture in the United States has seen widespread adoption, particularly in large-scale farming operations. The use of GPS, sensors, and data analytics has led to increased efficiency, reduced input costs, and higher yields.

Economic Outcomes

The economic benefits of precision agriculture include improved profitability, better resource management, and enhanced sustainability. However, the high initial investment and learning curve pose challenges for smaller farms.

Brazil: Soybean Production

Expansion and Challenges

Brazil is one of the world's largest producers of soybeans, driven by high global demand. The expansion of soybean production has brought economic benefits but also challenges related to deforestation, land use conflicts, and environmental degradation.

Policy Implications

Government policies supporting soybean production, such as subsidies and infrastructure development, have boosted the sector's growth. However, balancing economic growth with environmental conservation remains a critical challenge.

India: Smallholder Farming

Diversity and Resilience

India's agricultural landscape is characterized by diverse smallholder farms. These farms contribute significantly to food security and rural livelihoods but face challenges such as limited access to resources, technology, and markets.

Supportive Policies

Government policies in India, such as subsidies, credit schemes, and extension services, aim to support smallholder farmers. Initiatives to promote sustainable practices and improve market access are crucial for enhancing economic outcomes.

Future Directions and Innovations

Sustainable Agriculture

Agroecology

Agroecology integrates ecological principles into agricultural practices, promoting biodiversity, soil health, and sustainability. Policies supporting agroecological practices can enhance resilience and long-term economic viability.

Circular Economy

The circular economy approach in agriculture involves recycling waste, reducing inputs, and creating closed-loop systems. This approach can reduce costs, minimize environmental impact, and promote sustainability.

Digital Transformation

Smart Farming

Smart farming uses digital technologies to optimize agricultural practices. This includes IoT devices, remote sensing, and blockchain for traceability. Embracing digital transformation can lead to increased efficiency and profitability.

E-Commerce and Direct Sales

E-commerce platforms enable farmers to sell directly to consumers, bypassing traditional distribution channels. This approach can increase margins, expand market reach, and meet consumer demand for transparency and traceability.

Policy Innovations

Climate-Smart Policies

Climate-smart policies focus on reducing emissions, enhancing resilience, and promoting sustainable practices. These policies can provide incentives for adopting climate-smart agriculture and support research and innovation.

Inclusive Policies

Inclusive policies address the needs of marginalized groups, such as smallholder farmers, women, and indigenous communities. Ensuring equitable access to resources, technology, and markets is essential for sustainable economic development.

Conclusion

The economics of modern farming encompasses a wide range of factors, from technological advancements and market dynamics to policy frameworks and environmental challenges. By understanding these economic principles, stakeholders can make informed decisions to enhance productivity, profitability, and sustainability in the agricultural sector. Addressing the challenges and leveraging the opportunities presented by modern farming requires a collaborative and innovative approach, integrating economic, social, and environmental considerations. Through such efforts, the agricultural sector can contribute to global food security, rural development, and environmental conservation, ensuring a sustainable and prosperous future for farmers and communities worldwide.

References

Acheampong, E. O., Sloan, S., Sayer, J., & Macgregor, C. J. (2022). African forest-fringe farmers benefit from modern farming practices despite high environmental impacts. Land, 11(2), 145.

Kingwell, R. (2011). Managing complexity in modern farming. Australian Journal of Agricultural and Resource Economics, 55(1), 12-34.

Mishra, H., Tiwari, A. K., & Nishad, D. C. (2011). Economic Viability of Sustainable Agriculture Practices in Modern Farming. Advances in Agriculture Sciences Volume II, 24(4), 105.

Ram, N. (2020). Agriculture production & Modern Farming. International Journal of Research in Social Sciences, 10(2), 296-301.

Snyder, K. A., Sulle, E., Massay, D. A., Petro, A., Qamara, P., & Brockington, D. (2020). "Modern" farming and the transformation of livelihoods in rural Tanzania. Agriculture and Human Values, 37, 33-46.

Terziev, D. (2019). The family farm-an institution of modern farming. Agricultural Sciences/ Agrarni Nauki, 11(25).

Troughton, M. J. (2014). Farming systems in the modern world. In Progress in Agricultural Geography (Routledge Revivals) (pp. 93-123). Routledge.

Wang, H. H., Wang, Y., & Delgado, M. S. (2014). The transition to modern agriculture: Contract farming in developing economies. American Journal of Agricultural Economics, 96(5), 1257-1271.

12

Agroecology Harmonizing Agriculture and Ecology

Abstract

Agroecology is an integrated approach to farming that seeks to harmonize agricultural practices with ecological principles. It emphasizes biodiversity, sustainability, and the health of ecosystems, advocating for farming systems that work with nature rather than against it. This article explores the principles and practices of agroecology, its historical roots, and its contemporary applications. It examines the benefits and challenges of adopting agroecological methods, including case studies from various regions. The article also discusses the policy and social dimensions of agroecology, highlighting the role of community participation, knowledge sharing, and supportive policies in promoting agroecological transitions. By understanding and implementing agroecology, farmers and stakeholders can contribute to a more sustainable and resilient agricultural future.

Introduction

Agroecology represents a paradigm shift in the way we approach farming, focusing on the integration of agricultural practices with ecological principles. It aims to create sustainable farming systems that are environmentally sound, economically viable, and socially just. This holistic approach considers the complex interactions between plants, animals, humans, and the environment, promoting biodiversity, soil health, and ecosystem services. Agroecology is increasingly recognized as a viable alternative to conventional, industrial agriculture, which often relies on synthetic inputs and monocultures that can degrade the environment and compromise long-term sustainability.

Historical Roots of Agroecology

Traditional Farming Practices

Agroecology draws on traditional farming practices that have been developed and refined over centuries. Indigenous and peasant communities around the world have long practiced forms of agriculture that are inherently sustainable and resilient, relying on local knowledge, crop diversity, and ecological balance.

Emergence of Agroecology as a Science

The term "agroecology" emerged in the 20th century, combining the fields of agronomy and ecology. Early pioneers such as Miguel Altieri and Stephen Gliessman emphasized the need to study agricultural systems from an ecological perspective, integrating scientific knowledge with traditional practices.

Evolution into a Movement

Agroecology has evolved from a scientific discipline into a social movement advocating for food sovereignty, environmental justice, and sustainable development. It is supported by various grassroots organizations, non-governmental organizations (NGOs), and international institutions that promote agroecological principles and practices.

Principles of Agroecology

Biodiversity

Crop Diversity

Agroecology promotes the cultivation of diverse crops, including intercropping, crop rotation, and polycultures. Crop diversity enhances resilience to pests, diseases, and climate variability, and contributes to soil health and ecosystem stability.

Agroforestry

Agroforestry integrates trees and shrubs into agricultural landscapes, providing multiple benefits such as shade, windbreaks, soil enrichment, and habitat for wildlife. This practice enhances biodiversity and supports ecosystem services.

Soil Health

Organic Matter

Maintaining and enhancing soil organic matter is crucial for soil fertility and structure. Agroecological practices such as composting, cover cropping, and reduced tillage help increase organic matter and improve soil health.

Soil Erosion Control

Preventing soil erosion is essential for sustaining agricultural productivity. Techniques such as contour farming, terracing, and maintaining ground cover help protect soil from erosion and degradation.

Ecosystem Services

Pollination

Agroecological systems support pollinator populations by providing diverse habitats and reducing pesticide use. Pollinators are essential for the reproduction of many crops and maintaining biodiversity.

Pest and Disease Management

Agroecology emphasizes biological pest control and integrated pest management (IPM) to reduce reliance on synthetic pesticides. Natural predators, crop rotation, and habitat manipulation are used to manage pests and diseases sustainably.

Water Management

Efficient Water Use

Agroecological practices aim to optimize water use through techniques such as rainwater harvesting, drip irrigation, and maintaining soil moisture. Efficient water use is critical for sustaining agriculture in water-scarce regions.

Watershed Management

Watershed management involves protecting and restoring water sources and ecosystems. Agroecology promotes practices that enhance water infiltration, reduce runoff, and improve water quality.

Social and Economic Dimensions

Food Sovereignty

Agroecology supports food sovereignty, the right of communities to control their own food systems. It emphasizes local food production, equitable access to resources, and empowerment of smallholder farmers.

Community Participation

Community participation is central to agroecology. Collaborative approaches, knowledge sharing, and collective decision-making strengthen social cohesion and enhance the adoption of sustainable practices.

Benefits of Agroecology

Environmental Benefits

Biodiversity Conservation

Agroecology promotes biodiversity at all levels, from soil microorganisms to plant and animal species. This biodiversity enhances ecosystem resilience and provides essential services such as pollination and pest control.

Climate Change Mitigation

Agroecological practices contribute to climate change mitigation by sequestering carbon in soils and vegetation, reducing greenhouse gas emissions, and enhancing the resilience of farming systems to climate impacts.

Reduced Chemical Inputs

By minimizing the use of synthetic fertilizers and pesticides, agroecology reduces pollution, protects water quality, and enhances soil health. This approach also reduces the risk of pesticide resistance and health impacts on farmers and consumers.

Economic Benefits

Cost Savings

Agroecological practices can reduce costs associated with synthetic inputs, machinery, and irrigation. Farmers can save money by using locally available resources and traditional knowledge.

Diversified Income

Diversified farming systems provide multiple sources of income through a variety of crops, livestock, and value-added products. This diversification reduces economic risks and enhances financial stability.

Social Benefits

Improved Nutrition

Agroecology promotes diverse and nutritious diets by encouraging the production and consumption of a variety of crops and livestock. This diversity enhances food security and nutritional outcomes.

Empowerment of Farmers

Agroecology empowers farmers by valuing their knowledge, fostering innovation, and strengthening local food systems. It promotes autonomy and reduces dependence on external inputs and markets.

Challenges of Agroecology

Knowledge and Training

Knowledge Transfer

One of the challenges in adopting agroecology is the transfer of knowledge and skills. Farmers need access to training and education to implement agroecological practices effectively.

Research and Extension

Research and extension services need to support agroecology by developing context-specific solutions and disseminating information. Investment in agroecological research is essential for advancing the field.

Policy and Institutional Barriers

Policy Support

Lack of supportive policies can hinder the adoption of agroecology. Governments need to create enabling environments through policies that promote sustainable agriculture, provide incentives, and support smallholder farmers.

Institutional Frameworks

Institutional frameworks should facilitate the integration of agroecology into agricultural development programs. Collaboration among government agencies, NGOs, and farmers' organizations is crucial for success.

Market Access

Fair Trade

Access to fair and equitable markets is essential for the economic viability of agroecological farming. Fair trade initiatives and local food systems can help ensure that farmers receive fair prices for their products.

Certification and Standards

Certification schemes for organic and agroecological products can enhance market access but may also pose challenges due to costs and complexity. Simplified and accessible certification processes are needed.

Case Studies

Latin America: Agroecology and Food Sovereignty

Brazil

In Brazil, agroecology has been promoted by social movements such as the Landless Workers' Movement (MST) and supported by government policies. Agroecological farms have demonstrated increased productivity, resilience to climate change, and enhanced food security.

Cuba

Cuba's transition to agroecology was driven by necessity following the collapse of the Soviet Union and the loss of imported inputs. The country has developed a robust agroecological system that emphasizes local production, biodiversity, and sustainability.

Africa: Agroecology for Resilience

Senegal

In Senegal, agroecology has been promoted through farmer-to-farmer networks and training programs. Agroecological practices such as agroforestry, intercropping, and composting have improved soil fertility, increased yields, and enhanced resilience to climate change.

Kenya

Kenya has seen the growth of agroecological initiatives supported by NGOs and research institutions. These initiatives focus on soil health, water management, and crop diversification, leading to improved food security and livelihoods.

Europe: Agroecology in Action

France

In France, agroecology is supported by national policies and a strong movement of farmers and consumers. Practices such as organic farming, agroforestry, and direct sales to consumers have increased biodiversity, reduced chemical use, and strengthened local food systems.

Spain

Spain has a vibrant agroecology movement, with many farmers adopting sustainable practices and participating in local food networks. Agroecological practices have improved soil health, biodiversity, and community resilience.

Policy and Social Dimensions of Agroecology

Policy Frameworks

National Policies

National policies that support agroecology include subsidies for sustainable practices, research funding, and land reform. Governments can promote agroecology through comprehensive agricultural policies that integrate environmental and social goals.

International Agreements

International agreements, such as the United Nations' Sustainable Development Goals (SDGs), provide a framework for promoting agroecology. These agreements emphasize the importance of sustainable agriculture, food security, and environmental protection.

Social Movements

Grassroots Organizations

Grassroots organizations play a crucial role in advancing agroecology by advocating for farmers' rights, promoting sustainable practices, and building networks of support. These organizations empower farmers and facilitate knowledge exchange.

Participatory Approaches

Participatory approaches, such as farmer field schools and community-based research, engage farmers in the development and implementation of agroecological practices. These approaches enhance local knowledge and innovation.

Education and Training

Formal Education

Integrating agroecology into formal education programs, from primary schools to universities, is essential for building knowledge and skills. Educational institutions can promote agroecology through curricula, research, and extension services.

Informal Training

Informal training, such as workshops, field days, and farmer-to-farmer exchanges, provides practical knowledge and hands-on experience. These training programs are vital for disseminating agroecological practices.

Future Directions

Research and Innovation

Interdisciplinary Research

Interdisciplinary research that integrates agronomy, ecology, social sciences, and economics is crucial for advancing agroecology. Collaborative research efforts can develop innovative solutions and address complex challenges.

Technological Innovations

Technological innovations, such as digital tools, remote sensing, and biotechnologies, can support agroecology by enhancing monitoring, decision-making, and resource management. These technologies should be accessible and tailored to farmers' needs.

Scaling Up Agroecology

Policy Support

Scaling up agroecology requires supportive policies that promote sustainable agriculture, provide financial incentives, and remove barriers. Policymakers should prioritize agroecology in national and international agendas.

Community Engagement

Engaging communities in the promotion and adoption of agroecology is essential for scaling up. Community-based approaches that emphasize participation, knowledge sharing, and collective action can drive widespread change.

Global Collaboration

International Networks

International networks and collaborations can facilitate the exchange of knowledge, experiences, and best practices. Organizations such as the FAO and the International Panel of Experts on Sustainable Food Systems (IPES-Food) play a key role in promoting agroecology globally.

Solidarity and Advocacy

Solidarity among farmers, consumers, and advocates is crucial for advancing agroecology. Global advocacy efforts can raise awareness, influence policies, and mobilize resources for sustainable agriculture.

Conclusion

Agroecology offers a holistic and sustainable approach to agriculture that harmonizes farming practices with ecological principles. By promoting biodiversity, soil health, ecosystem services, and social equity, agroecology addresses the multifaceted challenges of modern agriculture. The benefits of agroecology extend beyond environmental sustainability to include economic viability and social well-being. However, realizing the full potential of agroecology requires overcoming challenges related to knowledge transfer, policy support, market access, and community engagement. Through collaborative efforts, innovative research, and supportive policies, agroecology can contribute to a more resilient, equitable, and sustainable agricultural future.

References

Cai, J., Li, X., Liu, L., Chen, Y., Wang, X., & Lu, S. (2021). Coupling and coordinated development of new urbanization and agro-ecological environment in China. Science of The Total Environment, 776, 145837.'

Fischer, G., Nachtergaele, F. O., Van Velthuizen, H., Chiozza, F., Francheschini, G., Henry, M., & Tramberend, S. (2021). Global agro-ecological zones (gaez v4)-model documentation.

Francis, C., Lieblein, G., Gliessman, S., Breland, T. A., Creamer, N., Harwood, R., & Poincelot, R. (2003). Agroecology: The ecology of food systems. Journal of sustainable agriculture, 22(3), 99-118.

Kerr, R. B., Madsen, S., Stüber, M., Liebert, J., Enloe, S., Borghino, N., ... & Wezel, A. (2021). Can agroecology improve food security and nutrition? A review. Global Food Security, 29, 100540.

Lescourret, F., Magda, D., Richard, G., Adam-Blondon, A. F., Bardy, M., Baudry, J., ... & Soussana, J. F. (2015). A social–ecological approach to managing multiple agro-ecosystem services. Current Opinion in Environmental Sustainability, 14, 68-75.

Méndez, V. E., Bacon, C. M., & Cohen, R. (2013). Agroecology as a transdisciplinary, participatory, and action-oriented approach. Agroecology and Sustainable Food Systems, 37(1), 3-18.

13

Livestock Farming Innovations and Challenges

Abstract

Livestock farming is a critical component of global agriculture, providing essential resources such as meat, milk, wool, and leather. This article explores the innovations and challenges facing the livestock farming industry. It discusses the latest technological advancements, such as precision livestock farming, genetic improvements, and sustainable practices, that are transforming the sector. Additionally, the article addresses the economic, environmental, and social challenges that farmers encounter, including market fluctuations, climate change, animal welfare, and disease management. By examining both the innovations and the obstacles, this article provides a comprehensive overview of the current state and future prospects of livestock farming, highlighting the need for sustainable and resilient farming practices.

Introduction

Livestock farming has been an integral part of human civilization for millennia, providing food, clothing, and labor. In recent decades, the industry has experienced significant changes due to technological advancements, evolving consumer preferences, and growing environmental concerns. This article aims to provide an in-depth analysis of the innovations and challenges in livestock farming, offering insights into how the industry can adapt and thrive in the future.

Technological Innovations in Livestock Farming

Precision Livestock Farming

Monitoring and Data Collection

Precision livestock farming (PLF) involves the use of advanced technologies to monitor and manage livestock more effectively. Sensors, GPS, and data analytics enable farmers to track animal health, behavior, and productivity in real-time. These technologies help optimize feeding, breeding, and overall herd management.

Automation and Robotics

Automation and robotics are increasingly used in livestock farming to perform tasks such as milking, feeding, and cleaning. Robotic milking systems, for example, can increase efficiency, reduce labor costs, and improve animal welfare by allowing cows to be milked on demand.

Genetic Improvements

Selective Breeding

Selective breeding has been used for centuries to enhance desirable traits in livestock. Modern techniques, such as genomic selection, allow for more precise and rapid improvements. This can lead to higher productivity, disease resistance, and better quality products.

Biotechnology

Biotechnological advancements, such as gene editing and cloning, offer new possibilities for improving livestock genetics. These technologies can potentially enhance growth rates, feed efficiency, and resistance to diseases, although they also raise ethical and regulatory concerns.

Sustainable Practices

Sustainable Feeding

Sustainable feeding practices aim to reduce the environmental impact of livestock farming by optimizing feed use and incorporating alternative feed sources. Innovations in feed formulation, such as using insect protein or algae, can improve efficiency and reduce greenhouse gas emissions.

Waste Management

Effective waste management is crucial for minimizing the environmental footprint of livestock farming. Technologies such as anaerobic digestion and composting convert animal waste into valuable resources like biogas and organic fertilizers, reducing pollution and greenhouse gas emissions.

Health and Welfare Innovations

Disease Prevention and Control

Advances in veterinary science and disease management, including vaccines, diagnostics, and biosecurity measures, play a vital role in maintaining animal health. Early detection and rapid response to disease outbreaks are essential for preventing widespread infections and economic losses.

Animal Welfare

Improving animal welfare is a growing concern for both consumers and producers. Innovations such as comfortable housing, enrichment activities, and stress-reducing handling techniques contribute to better living conditions and overall well-being of livestock.

Economic Challenges

Market Volatility

Price Fluctuations

Livestock farmers often face significant price volatility for their products, driven by factors such as supply and demand dynamics, trade policies, and market speculation. Price fluctuations can affect profitability and financial stability.

Global Trade

The global trade of livestock and livestock products exposes farmers to international market risks and competition. Trade agreements, tariffs, and non-tariff barriers can impact market access and profitability.

Access to Capital

Financing

Securing financing for livestock farming operations can be challenging, especially for small and medium-sized enterprises. High initial costs for technology adoption and infrastructure improvements require access to affordable credit and investment.

Risk Management

Effective risk management strategies, such as insurance and futures contracts, are essential for mitigating the financial risks associated with livestock farming. These tools can help farmers manage price volatility, disease outbreaks, and climate-related impacts.

Environmental Challenges

Climate Change

Impact on Livestock

Climate change poses significant challenges for livestock farming, including heat stress, altered disease patterns, and changes in feed availability. These impacts can reduce productivity and increase the vulnerability of livestock systems.

Mitigation Strategies

Adopting climate-smart practices, such as improved pasture management, efficient water use, and greenhouse gas mitigation technologies, is crucial for reducing the environmental footprint of livestock farming. These practices can enhance resilience and sustainability.

Land and Water Use

Resource Competition

Livestock farming competes with other land and water uses, such as crop production and urban development. Balancing these competing demands requires efficient resource management and sustainable land use practices.

Degradation and Pollution

Overgrazing, deforestation, and improper waste management can lead to land degradation, water pollution, and loss of biodiversity. Implementing sustainable grazing practices, riparian buffers, and waste recycling can mitigate these impacts.

Social Challenges

Animal Welfare Concerns

Public Perception

Growing awareness of animal welfare issues influences consumer preferences and regulatory policies. Farmers must adapt to these changing expectations by adopting humane practices and transparent communication.

Ethical Considerations

Ethical considerations regarding the treatment of animals, including the use of intensive farming systems and genetic modifications, pose challenges for the livestock industry. Balancing productivity with ethical practices is essential for maintaining social license to operate.

Labor and Workforce Issues

Labor Shortages

Labor shortages, particularly in rural areas, can hinder livestock farming operations. Automation and mechanization can alleviate some of these challenges, but skilled labor is still required for management and oversight.

Working Conditions

Improving working conditions for farmworkers is crucial for ensuring a sustainable and ethical livestock industry. This includes fair wages, safe working environments, and opportunities for training and development.

Case Studies of Innovative Practices

Dairy Farming in the Netherlands

Precision Farming Techniques

Dutch dairy farms have adopted precision farming techniques, including robotic milking systems and automated feed management. These innovations have led to increased efficiency, reduced labor costs, and improved animal welfare.

Sustainable Practices

The Netherlands is also a leader in sustainable dairy farming, with practices such as nutrient recycling, renewable energy use, and greenhouse gas mitigation. These efforts contribute to the overall sustainability and resilience of the dairy sector.

Beef Farming in Brazil

Integrated Crop-Livestock Systems

Brazilian beef farms are increasingly adopting integrated crop-livestock systems, which combine livestock production with crop cultivation. This approach enhances resource use efficiency, improves soil health, and reduces deforestation pressure.

Carbon-Neutral Beef

Initiatives to produce carbon-neutral beef in Brazil involve measures such as reforestation, improved pasture management, and methane reduction technologies. These efforts aim to reduce the carbon footprint of beef production and promote sustainability.

Poultry Farming in the United States

Antibiotic-Free Production

Consumer demand for antibiotic-free poultry has driven innovation in disease management and animal health. Practices such as improved biosecurity, vaccination, and alternative treatments help maintain productivity without relying on antibiotics.

Welfare Improvements

U.S. poultry farms are implementing welfare improvements, including enriched environments, humane handling practices, and transparent labeling. These changes address consumer concerns and enhance the overall quality of poultry products.

Policy and Regulatory Landscape

National Policies

Subsidies and Support Programs

National policies, such as subsidies and support programs, play a crucial role in supporting livestock farmers. These programs can provide financial assistance, promote research and innovation, and incentivize sustainable practices.

Environmental Regulations

Environmental regulations aimed at reducing pollution, conserving natural resources, and mitigating climate change impact livestock farming. Compliance with these regulations requires investment in sustainable practices and technologies.

International Frameworks

Trade Agreements

International trade agreements influence the competitiveness and market access of livestock products. Harmonizing standards and reducing trade barriers can enhance global trade and benefit livestock farmers.

Sustainable Development Goals

The United Nations' Sustainable Development Goals (SDGs) provide a framework for promoting sustainable livestock farming. These goals emphasize the importance of environmental sustainability, food security, and social equity.

Future Directions and Innovations

Technological Advancements

Smart Farming

The future of livestock farming lies in the continued development and adoption of smart farming technologies. Innovations such as artificial intelligence, blockchain, and IoT can enhance efficiency, traceability, and sustainability.

Alternative Proteins

The rise of alternative proteins, including plant-based and lab-grown meat, presents both challenges and opportunities for the livestock industry. These alternatives could reduce the environmental impact of meat production and meet changing consumer preferences.

Sustainable Practices

Regenerative Agriculture

Regenerative agriculture practices, such as holistic grazing and agroforestry, aim to restore soil health, enhance biodiversity, and sequester carbon. These practices offer a sustainable path forward for livestock farming.

Circular Economy

The circular economy approach in livestock farming involves recycling waste, reducing inputs, and creating closed-loop systems. This approach can minimize environmental impact and enhance resource efficiency.

Policy Innovations

Climate-Smart Policies

Climate-smart policies that promote resilience and sustainability are essential for the future of livestock farming. These policies should support research, provide incentives, and facilitate the adoption of climate-smart practices.

Inclusive Policies

Inclusive policies that address the needs of smallholder farmers, women, and marginalized communities are crucial for ensuring equitable and sustainable development. Access to resources, training, and markets should be prioritized.

Conclusion

Livestock farming is at a crossroads, facing significant challenges and opportunities in an era of rapid change. Technological innovations, sustainable practices, and supportive policies are essential for addressing these challenges and ensuring the future viability of the industry. By embracing these innovations and overcoming obstacles, livestock farmers can contribute to a sustainable, resilient, and equitable agricultural system that meets the needs of a growing global population.

References

Bernet, N., & Béline, F. (2009). Challenges and innovations on biological treatment of livestock effluents. Bioresource technology, 100(22), 5431-5436.

Džermeikaitė, K., Bačėninaitė, D., & Antanaitis, R. (2023). Innovations in cattle farming: application of innovative technologies and sensors in the diagnosis of diseases. Animals, 13(5), 780.

García-Martínez, A., Rivas-Rangel, J., Rangel-Quintos, J., Espinosa, J. A., Barba, C., & de-Pablos-Heredero, C. (2016). A methodological approach to evaluate livestock innovations on small-scale farms in developing countries. Future internet, 8(2), 25.

Moraine, M., Duru, M., Nicholas, P., Leterme, P., & Therond, O. (2014). Farming system design for innovative crop-livestock integration in Europe. Animal, 8(8), 1204-1217.

Papakonstantinou, G. I., Voulgarakis, N., Terzidou, G., Fotos, L., Giamouri, E., & Papatsiros, V. G. (2024). Precision Livestock Farming Technology: Applications and Challenges of Animal Welfare and Climate Change. Agriculture, 14(4), 620.

Subach, T. I., & Shmeleva, Z. N. (2022, December). Introduction of digital innovations in livestock farming. In IOP Conference Series: Earth and Environmental Science (Vol. 1112, No. 1, p. 012079). IOP Publishing.

14

Agri-business and Entrepreneurship in the 21st Century

Abstract

Agribusiness and entrepreneurship in the 21st century are evolving rapidly due to technological advancements, changing consumer preferences, and the increasing focus on sustainability. This article delves into the dynamic landscape of modern agribusiness, examining the innovations driving growth and the challenges that entrepreneurs face. It explores key trends such as digital agriculture, sustainable practices, and value chain integration, highlighting the role of startups and established companies in transforming the industry. The article also addresses the economic, social, and environmental impacts of agribusiness and provides insights into the future of agricultural entrepreneurship. By understanding these developments, stakeholders can better navigate the complexities of the contemporary agricultural sector and capitalize on emerging opportunities.

Introduction

The 21st century has brought significant changes to agribusiness, driven by technological innovation, globalization, and a growing emphasis on sustainability. Agribusiness encompasses all activities related to the production, processing, distribution, and marketing of agricultural products. Entrepreneurship plays a critical role in this sector, as new ventures and innovations address the evolving needs and challenges of modern agriculture. This article explores the current state of agribusiness and entrepreneurship, examining key trends, innovations, challenges, and future directions.

The Evolution of Agribusiness

Historical Context

Agribusiness has a long history, evolving from small-scale subsistence farming to large-scale industrial agriculture. The Green Revolution of the mid-20th century marked a significant shift, introducing high-yield crop varieties, chemical fertilizers, and advanced irrigation techniques that boosted agricultural productivity worldwide.

The Modern Agribusiness Landscape

Today, agribusiness is characterized by a complex global supply chain, advanced technologies, and diverse market demands. It includes various sectors such as crop production, livestock farming, agro-processing, and agricultural services. The integration of technology and innovation has transformed traditional farming practices, making agribusiness more efficient, productive, and sustainable.

Technological Innovations in Agribusiness

Digital Agriculture

Precision Farming

Precision farming involves using data-driven technologies to optimize agricultural practices. Tools such as GPS, sensors, drones, and satellite imagery provide real-time information on soil health, crop growth, and weather conditions. This data enables farmers to make informed decisions, reduce input costs, and increase yields.

Internet of Things (IoT)

IoT devices are revolutionizing agribusiness by connecting various farming equipment and systems. Smart sensors monitor soil moisture, temperature, and nutrient levels, enabling automated irrigation and fertilization. IoT technology also enhances livestock management by tracking animal health and behavior.

Biotechnology

Genetic Engineering

Biotechnology, including genetic engineering and CRISPR gene editing, is advancing crop and livestock breeding. These technologies enable the development of disease-resistant, drought-tolerant, and high-yielding varieties, improving productivity and sustainability.

Microbial Solutions

Microbial solutions, such as biofertilizers and biopesticides, leverage beneficial microorganisms to enhance soil health and protect crops. These sustainable alternatives to chemical inputs promote organic farming and reduce environmental impact.

Automation and Robotics

Autonomous Machinery

Automation and robotics are transforming labor-intensive agricultural tasks. Autonomous tractors, harvesters, and drones perform planting, weeding, and harvesting with high precision and efficiency, reducing labor costs and increasing productivity.

Robotic Milking Systems

In the dairy sector, robotic milking systems improve milking efficiency and animal welfare. These systems allow cows to be milked on demand, reducing stress and increasing milk yield.

Sustainable Practices in Agribusiness

Regenerative Agriculture

Soil Health

Regenerative agriculture focuses on restoring and maintaining soil health through practices such as cover cropping, no-till farming, and crop rotation. Healthy soils sequester carbon, retain water, and support biodiversity, enhancing the resilience and sustainability of farming systems.

Agroforestry

Agroforestry integrates trees and shrubs into agricultural landscapes, providing multiple benefits such as shade, windbreaks, and habitat for wildlife. This practice enhances biodiversity, improves soil health, and increases farm productivity.

Circular Economy

Waste Management

A circular economy approach in agribusiness involves recycling and reusing agricultural waste. Technologies such as anaerobic digestion convert organic waste into biogas and organic fertilizers, reducing waste and creating valuable by-products.

Resource Efficiency

Resource-efficient practices, such as water-saving irrigation systems and energy-efficient machinery, reduce the environmental footprint of agribusiness. These practices enhance sustainability and reduce operational costs.

Sustainable Supply Chains

Traceability and Transparency

Consumers increasingly demand transparency in the food supply chain. Technologies such as blockchain enable traceability, providing information on the origin, production methods, and environmental impact of agricultural products. This transparency builds consumer trust and supports sustainable practices.

Fair Trade

Fair trade initiatives ensure that farmers receive fair prices and improve their livelihoods. By promoting equitable and sustainable trade practices, fair trade supports smallholder farmers and enhances the social impact of agribusiness.

Entrepreneurship in Agribusiness

The Role of Startups

Innovation and Disruption

Startups play a crucial role in driving innovation and disruption in agribusiness. By developing new technologies, business models, and solutions, startups address emerging challenges and create new opportunities in the agricultural sector.

Access to Capital

Access to capital is essential for the growth and success of agribusiness startups. Venture capital, crowdfunding, and government grants provide funding for research, development, and market expansion.

Challenges for Agribusiness Entrepreneurs

Market Access

Market access is a significant challenge for agribusiness entrepreneurs, especially in developing regions. Infrastructure limitations, trade barriers, and market volatility can hinder the entry and expansion of new ventures.

Regulatory Compliance

Navigating complex regulatory environments is essential for agribusiness entrepreneurs. Compliance with food safety, environmental, and labor regulations requires understanding and adherence to national and international standards.

Support Systems for Entrepreneurs

Incubators and Accelerators

Incubators and accelerators provide support to agribusiness startups through mentorship, networking, and access to resources. These programs help entrepreneurs develop their ideas, build prototypes, and scale their businesses.

Knowledge Sharing and Collaboration

Collaboration and knowledge sharing among stakeholders, including farmers, researchers, and industry experts, are crucial for the success of agribusiness ventures. Platforms and networks facilitate the exchange of ideas, best practices, and innovations.

Economic, Social, and Environmental Impacts

Economic Impact

Job Creation

Agribusiness and entrepreneurship contribute to job creation, particularly in rural areas. The growth of agribusiness ventures generates employment opportunities in farming, processing, distribution, and retail.

Economic Growth

Agribusiness drives economic growth by contributing to GDP, export revenues, and rural development. The sector's expansion stimulates investment, infrastructure development, and technological innovation.

Social Impact

Food Security

Agribusiness plays a vital role in ensuring food security by increasing food production, improving supply chain efficiency, and reducing food waste. Innovations in agribusiness enhance the availability and accessibility of nutritious food.

Community Development

Agribusiness initiatives support community development by providing education, healthcare, and social services. Empowering local communities and smallholder farmers enhances social well-being and economic resilience.

Environmental Impact

Biodiversity Conservation

Sustainable agribusiness practices promote biodiversity conservation by preserving natural habitats, protecting endangered species, and enhancing ecosystem services. Biodiverse farming systems are more resilient to environmental changes.

Climate Change Mitigation

Agribusiness can contribute to climate change mitigation by adopting practices that reduce greenhouse gas emissions, sequester carbon, and enhance energy efficiency. Sustainable farming practices help combat global warming and promote environmental sustainability.

Case Studies of Successful Agribusiness Ventures

CropIn Technology Solutions

Digital Agriculture Solutions

CropIn Technology Solutions is an Indian agritech startup that provides digital agriculture solutions to farmers. Their platform offers farm management, crop monitoring, and data analytics, helping farmers increase productivity and profitability.

Impact and Growth

CropIn's solutions have positively impacted thousands of farmers by enhancing decision-making, reducing input costs, and increasing yields. The company's growth and success demonstrate the potential of digital agriculture in transforming agribusiness.

Beyond Meat

Plant-Based Protein

Beyond Meat is a pioneer in the development of plant-based protein alternatives. Their innovative products mimic the taste and texture of meat, providing sustainable and healthy options for consumers.

Market Disruption

Beyond Meat has disrupted the traditional meat industry by addressing environmental concerns and changing consumer preferences. Their success highlights the growing demand for sustainable and ethical food products.

Agrosmart

Climate-Smart Agriculture

Agrosmart, a Brazilian agritech company, provides climate-smart agriculture solutions through IoT and data analytics. Their platform helps farmers optimize water use, monitor climate conditions, and make informed decisions.

Social and Environmental Impact

Agrosmart's solutions enhance the resilience of farming systems to climate change, improve water management, and increase crop productivity. The company's work demonstrates the importance of innovation in addressing environmental challenges in agribusiness.

Policy and Regulatory Frameworks

National Policies

Agricultural Innovation Policies

National policies that support agricultural innovation, research, and development are essential for fostering entrepreneurship in agribusiness. Governments can provide funding, tax incentives, and infrastructure to support innovation.

Sustainable Agriculture Policies

Sustainable agriculture policies promote the adoption of environmentally friendly practices, such as organic farming, conservation tillage, and integrated pest management. These policies encourage sustainable development in agribusiness.

International Agreements

Trade Agreements

International trade agreements facilitate the exchange of agricultural products and technologies. Harmonizing standards and reducing trade barriers can enhance market access and competitiveness for agribusiness entrepreneurs.

Sustainable Development Goals

The United Nations' Sustainable Development Goals (SDGs) provide a framework for promoting sustainable and inclusive agribusiness. These goals emphasize the importance of food security, environmental sustainability, and social equity.

Future Directions in Agribusiness and Entrepreneurship

Technological Advancements

Artificial Intelligence and Machine Learning

Artificial intelligence (AI) and machine learning (ML) are poised to revolutionize agribusiness by enhancing data analytics, predictive modeling, and automation. These technologies can improve decision-making, optimize resource use, and increase productivity.

Blockchain Technology

Blockchain technology can enhance transparency and traceability in the food supply chain. By providing secure and immutable records, blockchain can build consumer trust, reduce fraud, and ensure food safety.

Sustainable Innovations

Alternative Proteins

The development of alternative proteins, such as lab-grown meat and insect-based protein, offers sustainable and ethical food options. These innovations can reduce the environmental impact of meat production and meet the growing demand for protein.

Climate-Resilient Crops

Research and development of climate-resilient crops are crucial for addressing the challenges of climate change. Genetically modified and selectively bred crops that withstand extreme weather conditions can enhance food security and sustainability.

Entrepreneurial Ecosystems

Ecosystem Support

Building robust entrepreneurial ecosystems that provide support, resources, and networks is essential for the success of agribusiness ventures. Governments, private sector, and academia can collaborate to create conducive environments for innovation.

Inclusive Growth

Promoting inclusive growth in agribusiness ensures that smallholder farmers, women, and marginalized communities benefit from entrepreneurial opportunities. Inclusive policies and programs can enhance social equity and economic resilience.

Conclusion

Agribusiness and entrepreneurship in the 21st century are characterized by rapid innovation, evolving consumer preferences, and a growing emphasis on sustainability. Technological advancements, sustainable practices, and supportive policies are transforming the agricultural sector, creating new opportunities and addressing emerging challenges. By embracing these developments, agribusiness entrepreneurs can contribute to a sustainable, resilient, and equitable food system that meets the needs of a growing global population. The future of agribusiness lies in the hands of innovative entrepreneurs who can navigate the complexities of the modern agricultural landscape and drive positive change.

References

Carayannis, E. G., Rozakis, S., & Grigoroudis, E. (2018). Agri-science to agri-business: The technology transfer dimension. The Journal of Technology Transfer, 43, 837-843.

Hans, V. (2008). Agri-Business and Rural Management in India-Issues and Challenges. Agri-Business and Rural Management in India-Issues and Challenges (September 21, 2008).

Lokanadhan, K. (2009). Innovations in agri-business management. New India Publishing.

Payumo, J. G., Lemgo, E. A., & Maredia, K. (2017). Transforming Sub-Saharan Africa's agriculture through agribusiness innovation. Global Journal of Agricultural Innovation, Research & Development, 4, 1-12.

Senthil Vinayagam, S. (1998). Entrepreneurial behaviour of agri-business operators in Kerala (Doctoral dissertation, Department of Agricultural Extension, College of Horticulture, Vellanikkara).

Singh, R., Khanna, V., Ramappa, K. B., & Kumari, T. (2023). Agribusiness and Entrepreneurship. In Trajectory of 75 years of Indian Agriculture after Independence (pp. 725-743). Singapore: Springer Nature Singapore.

Swarnalakshmi, R., and Nandhini, A. Agri-Business in India. Dr. Ngp Arts and Science College, 168.

15

Water Management in Agriculture Strategies for Conservation

Abstract

Water management in agriculture is a critical component of ensuring sustainable food production and environmental conservation. This article explores various strategies for conserving water in agricultural practices, examining both traditional methods and modern technological innovations. It addresses the challenges posed by climate change, population growth, and resource scarcity, highlighting the importance of efficient water use. Key strategies discussed include irrigation optimization, soil moisture management, rainwater harvesting, and the adoption of drought-resistant crops. The article also considers policy measures and community-based approaches that support water conservation. By implementing these strategies, the agricultural sector can contribute to sustainable water management and enhance food security.

Introduction

Water is a vital resource for agricultural production, yet its availability is increasingly threatened by factors such as climate change, population growth, and competing demands from other sectors. Effective water management in agriculture is essential for ensuring food security, protecting ecosystems, and supporting rural livelihoods. This article examines various strategies for conserving water in agriculture, exploring both traditional practices and modern innovations that enhance water use efficiency and sustainability.

The Importance of Water in Agriculture

Water Requirements for Crops

Crops require water for various physiological processes, including photosynthesis, nutrient uptake, and growth. Different crops have varying water needs, depending on factors such as their growth stage, climate, and soil type. Understanding these requirements is crucial for optimizing water use in agricultural practices.

Water Scarcity and Agriculture

Water scarcity poses a significant challenge to agriculture, particularly in arid and semi-arid regions. Factors such as over-extraction of groundwater, inefficient irrigation practices, and climate change exacerbate water scarcity, threatening agricultural productivity and food security.

Traditional Water Management Practices

Irrigation Techniques

Surface Irrigation

Surface irrigation, including methods such as flood irrigation and furrow irrigation, is one of the oldest and most common forms of irrigation. While simple and low-cost, surface irrigation can be inefficient and wasteful if not managed properly.

Subsurface Irrigation

Subsurface irrigation involves the application of water below the soil surface, directly to the root zone. This method reduces water loss through evaporation and runoff, enhancing water use efficiency.

Rainwater Harvesting

Rainwater harvesting is a traditional practice that involves collecting and storing rainwater for agricultural use. Techniques such as rooftop harvesting, ponds, and check dams can capture and store rainwater, providing a supplementary water source during dry periods.

Soil and Water Conservation

Terracing

Terracing involves shaping the land into stepped levels to reduce soil erosion and surface runoff. This practice enhances water infiltration and retention, improving soil moisture availability for crops.

Mulching

Mulching involves covering the soil surface with organic or inorganic materials to reduce evaporation, suppress weeds, and improve soil moisture retention. Mulching can significantly enhance water conservation in agricultural fields.

Modern Technological Innovations

Advanced Irrigation Systems

Drip Irrigation

Drip irrigation delivers water directly to the root zone of plants through a network of tubes, emitters, and valves. This method minimizes water loss through evaporation and runoff, significantly improving water use efficiency.

Sprinkler Irrigation

Sprinkler irrigation mimics natural rainfall by distributing water through a system of pipes and sprinklers. While more efficient than surface irrigation, sprinkler systems can still result in water loss if not properly managed.

Soil Moisture Sensors

Soil moisture sensors provide real-time data on soil moisture levels, helping farmers optimize irrigation schedules and reduce water use. These sensors can be integrated with automated irrigation systems to ensure precise water application.

Remote Sensing and GIS

Remote sensing and Geographic Information Systems (GIS) technologies enable the monitoring and management of water resources at a large scale. Satellite imagery and aerial surveys provide valuable data on soil moisture, crop health, and water availability, supporting informed decision-making.

Drought-Resistant Crops

Breeding and biotechnology have led to the development of drought-resistant crop varieties that can withstand water stress and maintain productivity under limited water conditions. These crops are crucial for improving agricultural resilience to water scarcity.

Integrated Water Management Strategies

Watershed Management

Watershed management involves the coordinated management of land and water resources within a watershed to achieve sustainable water use. Practices such as afforestation, soil conservation, and water harvesting enhance the health and functionality of watersheds, supporting agricultural water needs.

Conjunctive Use of Surface and Groundwater

Conjunctive use involves the combined use of surface water and groundwater resources to optimize water availability and sustainability. This approach balances the use of different water sources, reducing pressure on any single resource and enhancing overall water security.

Integrated Water Resource Management (IWRM)

IWRM is a comprehensive approach that promotes the coordinated management of water, land, and related resources to maximize economic and social benefits while ensuring environmental sustainability. IWRM principles guide the development of policies, institutions, and practices that support sustainable water management in agriculture.

Policy and Institutional Frameworks

Water Pricing and Incentives

Water pricing policies that reflect the true cost of water can encourage efficient water use and conservation. Incentives such as subsidies for water-saving technologies and practices can further promote sustainable water management in agriculture.

Water Rights and Allocation

Clear and equitable water rights and allocation systems are essential for managing water resources effectively. Policies that allocate water based on availability, demand, and sustainability criteria can prevent over-extraction and ensure fair access for all users.

Capacity Building and Education

Training and education programs that build the capacity of farmers, water managers, and policymakers are crucial for promoting sustainable water management practices. Knowledge sharing and extension services can disseminate best practices and innovations in water conservation.

Community-Based Approaches

Participatory Irrigation Management

Participatory Irrigation Management (PIM) involves the active involvement of farmers in the planning, operation, and maintenance of irrigation systems. PIM empowers communities to manage their water resources more effectively, enhancing efficiency and sustainability.

Water User Associations

Water User Associations (WUAs) are community-based organizations that manage local water resources and irrigation systems. WUAs facilitate collective action, promote equitable water distribution, and support the adoption of water-saving practices.

Community-Led Conservation Projects

Community-led conservation projects engage local communities in activities such as watershed restoration, rainwater harvesting, and soil conservation. These projects foster a sense of ownership and responsibility, leading to more sustainable water management outcomes.

Economic and Environmental Impacts

Economic Benefits

Increased Crop Yields

Efficient water management practices can increase crop yields by ensuring adequate and timely water supply. Higher yields translate into greater income and food security for farmers.

Reduced Costs

Water-saving technologies and practices can reduce input costs for farmers by minimizing water use, energy consumption, and labor requirements. These savings can improve farm profitability and sustainability.

Environmental Benefits

Reduced Water Extraction

Conserving water in agriculture reduces the pressure on freshwater resources, helping to maintain ecological balance and sustain aquatic ecosystems.

Enhanced Soil Health

Sustainable water management practices, such as soil moisture management and conservation tillage, improve soil health and structure. Healthy soils retain water more effectively and support diverse plant and microbial life.

Case Studies of Successful Water Management

Israel's Agricultural Water Management

Drip Irrigation Pioneers

Israel is a global leader in drip irrigation technology, which has revolutionized water management in agriculture. The widespread adoption of drip irrigation has significantly improved water use efficiency and agricultural productivity.

National Water Management Strategies

Israel's national water management strategies, including the desalination of seawater and the reuse of treated wastewater, have ensured a stable and sustainable water supply for agriculture and other sectors.

Water Conservation in India's Gujarat State

Community-Led Water Management

In Gujarat, community-led water management initiatives, such as the Sardar Patel Participatory Water Conservation Programme, have successfully mobilized local communities to conserve water and enhance agricultural productivity.

Check Dams and Rainwater Harvesting

The construction of check dams and the promotion of rainwater harvesting have significantly improved water availability in Gujarat's arid regions, supporting sustainable agriculture and rural livelihoods.

Future Directions in Agricultural Water Management

Climate-Smart Agriculture

Climate-smart agriculture practices that enhance resilience to climate change are crucial for sustainable water management. These practices include the adoption of drought-resistant crops, efficient irrigation systems, and integrated water resource management.

Innovations in Water-Saving Technologies

Ongoing research and development in water-saving technologies, such as advanced irrigation systems, soil moisture sensors, and remote sensing, will continue to improve water use efficiency and sustainability in agriculture.

Policy and Institutional Reforms

Policy and institutional reforms that promote sustainable water management, such as water pricing, rights allocation, and capacity building, are essential for addressing the challenges of water scarcity and ensuring long-term water security.

Conclusion

Water management in agriculture is a complex and multifaceted challenge that requires a combination of traditional practices, modern technologies, and innovative strategies. By adopting efficient irrigation methods, sustainable practices, and integrated water management approaches, the agricultural sector can enhance water use efficiency and contribute to sustainable water management. Policies and community-based initiatives that support water conservation are crucial for ensuring the availability of this vital resource for future generations. Through concerted efforts and collaboration, stakeholders can address the challenges of water scarcity and build a more resilient and sustainable agricultural system.

References

Gobarah, M. E., Tawfik, M. M., Thalooth, A. T., & Housini, E. A. E. (2015). Water conservation practices in agriculture to cope with water scarcity. International Journal of Water Resources and Arid Environments, 4(1), 20-29.

Kassam, A., Derpsch, R., & Friedrich, T. (2014). Global achievements in soil and water conservation: The case of Conservation Agriculture. International Soil and Water Conservation Research, 2(1), 5-13.

Qadir, M., Boers, T. M., Schubert, S., Ghafoor, A., & Murtaza, G. (2003). Agricultural water management in water-starved countries: challenges and opportunities. Agricultural water management, 62(3), 165-185.

Rastogi, M., Kolur, S. M., Burud, A., Sadineni, T., Sekhar, M., Kumar, R., & Rajput, A. (2024). Advancing Water Conservation Techniques in Agriculture for Sustainable Resource Management: A review. Journal of Geography, Environment and Earth Science International, 28(3), 41-53.

Sarvade, S., Upadhyay, V. B., Kumar, M., & Imran Khan, M. (2019). Soil and water conservation techniques for sustainable agriculture. Sustainable agriculture, forest and environmental management, 133-188.

Schaible, G., & Aillery, M. (2012). Water conservation in irrigated agriculture: Trends and challenges in the face of emerging demands. USDA-ERS Economic Information Bulletin, (99).

16

Soil Health and Management The Foundation of Agriculture

Abstract

Soil health and management are fundamental to sustainable agriculture, impacting crop productivity, environmental quality, and agricultural resilience. This article explores the various aspects of soil health, including its physical, chemical, and biological properties, and discusses management practices that enhance soil fertility and sustainability. The importance of organic matter, soil biodiversity, and nutrient cycling is highlighted, along with the challenges posed by soil degradation, erosion, and climate change. Through a detailed examination of traditional and innovative soil management strategies, such as crop rotation, cover cropping, conservation tillage, and soil amendments, the article provides comprehensive insights into maintaining and improving soil health. By understanding and implementing these practices, farmers and stakeholders can ensure long-term agricultural productivity and environmental stewardship.

Introduction

Soil health is the cornerstone of agricultural productivity and environmental sustainability. It encompasses the soil's ability to function as a living ecosystem, supporting plant and animal life, regulating water and nutrient cycles, and mitigating environmental impacts. Effective soil management practices are essential for maintaining and enhancing soil health, ensuring the sustainability of agricultural systems. This article delves into the components of soil health, examines the challenges of soil degradation, and explores various management practices that promote soil sustainability.

The Components of Soil Health

Physical Properties

Soil Texture

Soil texture, determined by the proportion of sand, silt, and clay particles, influences water retention, drainage, and root penetration. Ideal soil texture varies depending on the crop but generally, loamy soils are preferred for their balanced properties.

Soil Structure

Soil structure refers to the arrangement of soil particles into aggregates or clumps. Good soil structure enhances aeration, water infiltration, and root growth, while poor structure can lead to compaction and reduced productivity.

Chemical Properties

Soil pH

Soil pH affects nutrient availability and microbial activity. Most crops prefer a slightly acidic to neutral pH (6.0-7.5), although some plants have specific pH requirements.

Nutrient Content

Essential nutrients for plant growth include macronutrients (nitrogen, phosphorus, potassium) and micronutrients (zinc, copper, manganese). Soil tests can determine nutrient levels and guide fertilization practices.

Biological Properties

Soil Microorganisms

Soil health depends significantly on the diversity and activity of soil microorganisms, including bacteria, fungi, and protozoa. These organisms play crucial roles in decomposing organic matter, cycling nutrients, and suppressing soil-borne diseases.

Soil Organic Matter

Soil organic matter (SOM) consists of decomposed plant and animal residues. SOM improves soil structure, water retention, nutrient availability, and supports diverse soil life.

Challenges to Soil Health

Soil Degradation

Erosion

Soil erosion, caused by wind and water, removes the topsoil, which is rich in nutrients and organic matter. Erosion reduces soil fertility and can lead to desertification.

Compaction

Soil compaction from heavy machinery or livestock trampling reduces pore space, hindering root growth, water infiltration, and air exchange.

Loss of Organic Matter

Intensive farming practices, such as monocropping and excessive tillage, can deplete soil organic matter. Reduced SOM affects soil structure, water retention, and nutrient cycling.

Soil Contamination

Soil contamination from pesticides, heavy metals, and other pollutants can harm soil health and crop safety. Contaminated soils may require remediation to restore productivity.

Climate Change

Climate change affects soil health through altered precipitation patterns, increased temperatures, and extreme weather events. These changes can exacerbate soil erosion, degradation, and loss of biodiversity.

Soil Health Management Practices

Crop Rotation

Crop rotation involves growing different crops in a sequential manner on the same field. This practice prevents the buildup of pests and diseases, improves soil structure, and enhances nutrient cycling.

Cover Cropping

Cover crops, such as legumes and grasses, are grown between main crops to protect and enrich the soil. They reduce erosion, add organic matter, and enhance soil fertility through nitrogen fixation and other processes.

Conservation Tillage

Conservation tillage minimizes soil disturbance, preserving soil structure and organic matter. Practices include no-till, strip-till, and reduced-till farming, which also reduce erosion and improve water infiltration.

Organic Amendments

Organic amendments, such as compost, manure, and biochar, add organic matter and nutrients to the soil. These amendments improve soil structure, water retention, and microbial activity.

Integrated Pest Management (IPM)

IPM combines biological, cultural, mechanical, and chemical control methods to manage pests and diseases sustainably. This approach reduces reliance on chemical pesticides, protecting soil health and biodiversity.

Agroforestry

Agroforestry integrates trees and shrubs into agricultural systems. Trees provide shade, windbreaks, and organic matter, enhancing soil health and productivity.

Soil Testing and Monitoring

Regular soil testing and monitoring guide soil management decisions. Soil tests provide information on pH, nutrient levels, organic matter content, and other properties, helping farmers optimize soil fertility and health.

Innovative Soil Management Strategies

Precision Agriculture

Precision agriculture uses technology, such as GPS, sensors, and data analytics, to optimize soil and crop management. This approach enhances efficiency, reduces input use, and improves soil health.

Biofertilizers and Biostimulants

Biofertilizers contain living microorganisms that enhance nutrient availability and soil health. Biostimulants include natural substances that promote plant growth and soil microbial activity.

Soil Microbiome Engineering

Soil microbiome engineering involves manipulating soil microbial communities to improve soil health and crop productivity. Techniques include inoculating soils with beneficial microbes and managing soil conditions to favor desirable microorganisms.

Conservation Agriculture

Conservation agriculture integrates principles of minimal soil disturbance, permanent soil cover, and crop diversification. This holistic approach improves soil health, water management, and resilience to climate change.

Policy and Institutional Support

Sustainable Agriculture Policies

Governments and institutions can promote soil health through policies that support sustainable agriculture. Incentives, subsidies, and regulations can encourage practices such as cover cropping, conservation tillage, and organic farming.

Research and Extension Services

Research and extension services provide farmers with knowledge, tools, and technologies for improving soil health. Collaborative efforts between researchers, extension agents, and farmers are essential for disseminating best practices.

Community-Based Initiatives

Community-based initiatives, such as farmer cooperatives and local conservation projects, foster collective action for soil health. These initiatives facilitate knowledge sharing, resource pooling, and implementation of sustainable practices.

Case Studies of Successful Soil Management

The Loess Plateau, China

Rehabilitation Project

The Loess Plateau Rehabilitation Project transformed a degraded landscape into productive farmland through soil conservation, reforestation, and sustainable farming practices. The project improved soil health, reduced erosion, and enhanced livelihoods.

The Great Plains, USA

No-Till Farming

No-till farming practices in the Great Plains have significantly reduced soil erosion, improved water retention, and increased organic matter. Farmers adopting no-till methods have seen long-term benefits for soil health and productivity.

Sub-Saharan Africa

Agroforestry Systems

Agroforestry systems in Sub-Saharan Africa integrate trees with crops and livestock, enhancing soil fertility, reducing erosion, and providing multiple income sources for farmers. These systems are vital for sustainable agriculture in the region.

Future Directions in Soil Health Management

Climate Resilient Practices

Developing and adopting climate-resilient soil management practices is crucial for adapting to climate change. Practices such as drought-resistant crops, water conservation, and soil carbon sequestration enhance resilience.

Technological Advancements

Advancements in technology, such as remote sensing, drones, and AI, will continue to revolutionize soil health management. These tools provide precise, real-time data for informed decision-making.

Policy Integration

Integrating soil health into broader agricultural and environmental policies ensures a holistic approach to sustainability. Policies should address soil conservation, water management, biodiversity, and climate change mitigation.

Global Collaboration

Global collaboration and knowledge exchange are essential for addressing soil health challenges. International initiatives, such as the Global Soil Partnership, promote sustainable soil management practices worldwide.

Conclusion

Soil health and management are foundational to sustainable agriculture and environmental stewardship. By understanding the components of soil health and implementing effective management practices, farmers can enhance soil fertility, productivity, and resilience. Traditional methods, modern innovations, and supportive policies all play critical roles in maintaining and improving soil health. As the agricultural sector faces increasing challenges from soil degradation, climate change, and resource scarcity, a concerted effort is required to promote sustainable soil management practices. Through collaboration and innovation, stakeholders can ensure the long-term health of soils and the sustainability of agricultural systems.

References

Gobarah, M. E., Tawfik, M. M., Thalooth, A. T., & Housini, E. A. E. (2015). Water conservation practices in agriculture to cope with water scarcity. International Journal of Water Resources and Arid Environments, 4(1), 20-29.

Kassam, A., Derpsch, R., & Friedrich, T. (2014). Global achievements in soil and water conservation: The case of Conservation Agriculture. International Soil and Water Conservation Research, 2(1), 5-13.

Qadir, M., Boers, T. M., Schubert, S., Ghafoor, A., & Murtaza, G. (2003). Agricultural water management in water-starved countries: challenges and opportunities. Agricultural water management, 62(3), 165-185.

Rastogi, M., Kolur, S. M., Burud, A., Sadineni, T., Sekhar, M., Kumar, R., & Rajput, A. (2024). Advancing Water Conservation Techniques in Agriculture for Sustainable Resource Management: A review. Journal of Geography, Environment and Earth Science International, 28(3), 41-53.

Sarvade, S., Upadhyay, V. B., Kumar, M., & Imran Khan, M. (2019). Soil and water conservation techniques for sustainable agriculture. Sustainable agriculture, forest and environmental management, 133-188.

Schaible, G., & Aillery, M. (2012). Water conservation in irrigated agriculture: Trends and challenges in the face of emerging demands. USDA-ERS Economic Information Bulletin, (99).

17

Agricultural Education and Extension Services

Abstract

Agricultural education and extension services are crucial for the development and dissemination of knowledge and technologies that enhance agricultural productivity, sustainability, and resilience. This article explores the evolution, significance, and impact of agricultural education and extension services, highlighting their role in empowering farmers, improving food security, and fostering rural development. It examines the various models and approaches to agricultural education and extension, the challenges they face, and the innovative strategies being implemented to address these challenges. The article also discusses the integration of modern technologies and the importance of policy support in strengthening agricultural education and extension systems. By understanding and enhancing these services, stakeholders can ensure that farmers are well-equipped to meet the demands of modern agriculture and contribute to sustainable development.

Introduction

Agricultural education and extension services are foundational elements of agricultural development. They provide farmers with the knowledge, skills, and technologies necessary to improve agricultural practices, enhance productivity, and ensure sustainability. This article delves into the history, significance, and various aspects of agricultural education and extension services, exploring their role in modern agriculture and the challenges they face.

The Evolution of Agricultural Education and Extension Services

Historical Background

Agricultural education and extension services have evolved significantly over the centuries. Early agricultural education was often informal, passed down through generations of farmers. The establishment of agricultural schools and universities in the 19th and 20th centuries marked a formalization of agricultural education, providing structured curricula and research-based knowledge.

The Morrill Act and Land-Grant Universities

In the United States, the Morrill Act of 1862 established land-grant universities, which played a pivotal role in advancing agricultural education and research. These institutions were tasked with providing practical education in agriculture and the mechanical arts, making higher education accessible to a broader population.

The Rise of Extension Services

The Smith-Lever Act of 1914 established the Cooperative Extension Service in the United States, linking land-grant universities with farmers through county-based agents. This model of extension services, which provides practical, research-based information directly to farmers, has been adopted and adapted by many countries worldwide.

The Importance of Agricultural Education and Extension Services

Enhancing Agricultural Productivity

Agricultural education and extension services provide farmers with the latest knowledge and technologies, helping them adopt best practices and improve productivity. This includes information on crop management, soil health, pest control, and sustainable farming practices.

Promoting Sustainable Agriculture

Sustainability is a key focus of modern agricultural education and extension. These services promote practices that conserve resources, protect the environment, and ensure the long-term viability of agriculture. Topics such as organic farming, agroforestry, and integrated pest management are commonly covered.

Supporting Rural Development

Agricultural education and extension services play a crucial role in rural development by empowering farmers with the skills and knowledge needed to improve their livelihoods. They also support community development initiatives and help build social capital in rural areas.

Enhancing Food Security

By improving agricultural productivity and sustainability, agricultural education and extension services contribute to food security. They help farmers increase their yields, diversify their crops, and reduce post-harvest losses, ensuring a stable and sufficient food supply.

Models and Approaches to Agricultural Education and Extension

The Training and Visit (T&V) System

The T&V system, developed by the World Bank, is a top-down approach where extension agents visit farmers on a regular schedule to provide training and technical advice. While effective in delivering information, this model has been criticized for its lack of farmer participation and adaptability.

Farmer Field Schools (FFS)

The FFS approach is a participatory model where farmers learn through hands-on experience and experimentation. Farmers meet regularly to observe, discuss, and practice new techniques, fostering peer learning and empowerment.

Integrated Agricultural Research and Development (IAR4D)

The IAR4D approach integrates research, extension, and development activities, involving multiple stakeholders in a collaborative process. This model emphasizes the co-creation of knowledge and solutions, addressing complex agricultural challenges.

E-extension

E-extension leverages digital technologies to deliver information and services to farmers. This includes online platforms, mobile apps, and social media, which provide real-time access to information and facilitate communication between farmers and extension agents.

Challenges in Agricultural Education and Extension

Limited Resources and Funding

Many extension services, particularly in developing countries, face challenges related to limited resources and funding. This affects their ability to reach farmers, provide training, and adopt new technologies.

Capacity Building and Training

Extension agents often require ongoing training and capacity building to stay updated with the latest agricultural research and technologies. Inadequate training can limit their effectiveness and the quality of services they provide.

Gender Inequality

Gender inequality can restrict women's access to agricultural education and extension services. Addressing these disparities is crucial for empowering women farmers and enhancing overall agricultural productivity.

Climate Change

Climate change poses significant challenges for agriculture, requiring extension services to provide relevant and timely information on adaptation and mitigation strategies. This includes information on climate-resilient crops, water management, and sustainable farming practices.

Integration of Modern Technologies

While modern technologies offer significant opportunities for improving agricultural education and extension, integrating these technologies into existing systems can be challenging. This includes issues related to infrastructure, digital literacy, and access to technology.

Innovative Strategies in Agricultural Education and Extension

Mobile Technology and Digital Platforms

Mobile technology and digital platforms have revolutionized agricultural extension, providing farmers with real-time information and access to services. Examples include SMS-based advisories, mobile apps for pest identification, and online training modules.

Public-Private Partnerships

Public-private partnerships can enhance the reach and effectiveness of extension services. Private sector involvement can provide additional resources, expertise, and innovative solutions, complementing public extension efforts.

Farmer-to-Farmer Extension

Farmer-to-farmer extension leverages the knowledge and experience of lead farmers to disseminate information and practices within their communities. This approach fosters peer learning and ensures that information is contextually relevant.

Climate-Smart Agriculture

Climate-smart agriculture practices are being integrated into extension services to help farmers adapt to and mitigate the impacts of climate change. This includes promoting climate-resilient crops, conservation agriculture, and sustainable water management practices.

Participatory Research and Extension

Participatory research and extension involve farmers in the research process, ensuring that solutions are practical and relevant to their needs. This approach fosters a sense of ownership and enhances the adoption of new practices.

Policy Support for Agricultural Education and Extension

National Agricultural Policies

Supportive national agricultural policies are essential for the development and effectiveness of agricultural education and extension services. Policies should prioritize funding, capacity building, and the integration of modern technologies.

International Cooperation

International cooperation and partnerships can enhance the exchange of knowledge, technologies, and best practices in agricultural education and extension. Organizations such as the FAO, World Bank, and IFAD play significant roles in supporting these initiatives.

Gender-Inclusive Policies

Policies that promote gender equality in agricultural education and extension are crucial for ensuring that women farmers have equal access to services and opportunities. This includes targeted programs, capacity building, and addressing socio-cultural barriers.

Sustainable Development Goals (SDGs)

The United Nations Sustainable Development Goals (SDGs) provide a framework for promoting sustainable agriculture and rural development. Agricultural education and extension services play a vital role in achieving goals related to zero hunger, quality education, gender equality, and climate action.

Case Studies of Successful Agricultural Education and Extension Programs

India: Krishi Vigyan Kendras (KVKs)

The KVKs are a network of agricultural extension centers in India that provide training, demonstrations, and technical support to farmers. They have played a significant role in disseminating knowledge and improving agricultural practices across the country.

Kenya: Digital Green

Digital Green is a non-profit organization that uses video-based training to improve agricultural extension in Kenya. Farmers create and share videos on best practices, facilitating peer learning and the adoption of new techniques.

Brazil: EMBRAPA

The Brazilian Agricultural Research Corporation (EMBRAPA) integrates research, education, and extension services to support agricultural development. EMBRAPA's innovative approaches and technologies have significantly contributed to Brazil's agricultural productivity and sustainability.

Bangladesh: Agricultural Extension Services

Bangladesh's agricultural extension services have successfully implemented the FFS approach, empowering farmers through participatory learning and experimentation. This has led to improved productivity, food security, and resilience to climate change.

Future Directions in Agricultural Education and Extension

Embracing Digital Transformation

The digital transformation of agricultural education and extension services is essential for reaching a larger number of farmers and providing timely, relevant information. Future efforts should focus on expanding digital infrastructure, improving digital literacy, and developing user-friendly platforms.

Strengthening Collaboration

Collaboration between governments, research institutions, private sector, and civil society is crucial for enhancing agricultural education and extension services. Multi-stakeholder partnerships can leverage resources, expertise, and innovation to address complex agricultural challenges.

Fostering Innovation

Encouraging innovation in agricultural education and extension services is essential for addressing emerging challenges and opportunities. This includes developing new training methods, leveraging data analytics, and promoting sustainable practices.

Enhancing Capacity Building

Continuous capacity building for extension agents and farmers is vital for the effectiveness of agricultural education and extension services. Training programs should focus on current agricultural trends, technologies, and sustainable practices.

Promoting Inclusive Approaches

Inclusive approaches that address the needs of all farmers, including women, youth, and marginalized groups, are essential for equitable agricultural development. Policies and programs should aim to reduce barriers and ensure equal access to education and extension services.

Conclusion

Agricultural education and extension services are pivotal for the development and sustainability of agriculture. By providing farmers with the knowledge, skills, and technologies needed to improve their practices, these services enhance productivity, food security, and rural development. The integration of modern technologies, innovative strategies, and supportive policies is crucial for addressing the challenges faced by agricultural education and extension services. Through collaboration and continuous improvement, stakeholders can ensure that farmers are well-equipped to meet the demands of modern agriculture and contribute to sustainable development.

References

Charatsari, C., Papadaki-Klavdianou, A., & Michailidis, A. (2011). Farmers as consumers of agricultural education services: Willingness to pay and spend time. The Journal of Agricultural Education and Extension, 17(3), 253-266.

Klerkx, L. (2020). Advisory services and transformation, plurality and disruption of agriculture and food systems: towards a new research agenda for agricultural education and extension studies. The Journal of Agricultural Education and Extension, 26(2), 131-140.

Kosior, K. (2017, July). Agricultural education and extension in the age of Big Data. In European Seminar on Extension and Education.

Laurent, C., Cerf, M., & Labarthe, P. (2006). Agricultural extension services and market regulation: learning from a comparison of six EU countries. Journal of Agricultural Education and Extension, 12(1), 5-16.

Mulder, M., & Pachuau, A. (2011). How agricultural is agricultural education and extension?. The Journal of Agricultural Education and Extension, 17(3), 219-222.

Oloruntoba, A., & Adegbite, D. A. (2006). Improving agricultural extension services through university outreach initiatives: a case of farmers in model villages in Ogun State, Nigeria. Journal of agricultural education and extension, 12(4), 273-283.

18

The Future of Genetically Modified Organisms (GMOs)

Abstract

The future of genetically modified organisms (GMOs) presents a complex and multifaceted landscape, intertwining scientific innovation, regulatory challenges, environmental considerations, and societal acceptance. This article explores the potential advancements in GMO technology, including enhanced nutritional content, increased crop yields, and resistance to pests and diseases. It also examines the ethical and ecological implications, such as gene flow to non-GMO species and the potential impacts on biodiversity. Regulatory frameworks and public perception are discussed, highlighting the need for transparent communication and robust safety assessments. Ultimately, the future of GMOs will depend on a balanced approach that addresses both the opportunities and challenges, ensuring that this technology can contribute to sustainable agriculture and food security in a responsible manner.

Introduction

Genetically Modified Organisms (GMOs) have been at the forefront of agricultural biotechnology since the first genetically engineered crops were commercialized in the mid-1990s. These organisms, whose genetic material has been altered using modern genetic engineering techniques, offer potential solutions to some of the most pressing challenges in agriculture, including food security, environmental sustainability, and climate change. As we look towards the future, it is essential to consider both the technological advancements and the broader implications of GMOs on society and the environment. This article aims to provide a comprehensive overview of the future of GMOs, addressing key developments, challenges, and opportunities that lie ahead.

Technological Advancements in GMOs

New Techniques in Genetic Engineering

The advent of CRISPR-Cas9 and other gene-editing technologies has revolutionized the field of genetic engineering. Unlike traditional methods, which often involve inserting foreign DNA into an organism, these new techniques allow for precise modifications at specific locations in the genome. This precision reduces the risk of unintended effects and opens up new possibilities for creating GMOs with enhanced traits. Gene drives, another emerging technology, have the potential to spread desirable genes through wild populations, offering solutions to issues such as invasive species and vector-borne diseases.

Development of Nutrient-Enriched Crops

Biofortification, the process of increasing the nutritional value of crops through genetic modification, holds promise for addressing malnutrition in developing countries. Examples include Golden Rice, engineered to produce provitamin A, and iron-enriched beans. These crops have the potential to improve public health outcomes by providing essential nutrients that are often lacking in the diets of vulnerable populations.

Pest and Disease Resistance

One of the earliest and most successful applications of GMOs has been in developing crops resistant to pests and diseases. Bt crops, which produce a toxin derived from the bacterium Bacillus thuringiensis, have significantly reduced the need for chemical pesticides. Future advancements may include crops engineered to resist a broader range of pests and pathogens, reducing losses and improving yields.

Climate Resilience

As climate change poses increasing threats to agricultural productivity, developing crops that can withstand extreme weather conditions becomes crucial. Research is ongoing to create drought-tolerant, heat-resistant, and salinity-tolerant crops. These advancements will be essential for maintaining food production in regions most affected by climate change.

Future Crops: From Concept to Reality

The next generation of GMOs may include crops with entirely new capabilities, such as nitrogen-fixing cereals that reduce the need for synthetic fertilizers or plants engineered to sequester more carbon in the soil, contributing to climate mitigation efforts. These innovative applications have the potential to transform agriculture and contribute to environmental sustainability.

Environmental Impacts and Considerations

Gene Flow and Its Ecological Consequences

Gene flow, the transfer of genes from GMOs to wild relatives or non-GMO crops, raises ecological concerns. While gene flow can occur naturally, the introduction of genetically modified traits into wild populations may have unintended consequences, such as the creation of "superweeds" resistant to herbicides. Understanding and mitigating these risks is crucial for the responsible deployment of GMOs.

Effects on Biodiversity

The impact of GMOs on biodiversity is a subject of ongoing debate. While GMOs can contribute to biodiversity by reducing the need for chemical inputs and preserving natural habitats, there are concerns about their potential to outcompete or disrupt existing species. Comprehensive ecological studies are needed to assess these impacts and guide sustainable practices.

GMOs and Pesticide Use: Trends and Implications

The introduction of herbicide-resistant GMOs has led to increased use of certain herbicides, raising concerns about environmental and health impacts. Conversely, insect-resistant GMOs have reduced the need for chemical insecticides. Future developments should aim to balance these trends, promoting integrated pest management strategies that minimize negative effects.

Soil Health and GMO Cultivation

The long-term effects of GMO cultivation on soil health are not fully understood. Some studies suggest that GMOs can improve soil health by reducing the need for tillage and chemical inputs, while others highlight potential risks associated with altered microbial communities. Ongoing research is needed to fully understand these interactions and develop best practices.

Case Studies of Environmental Impacts

Examining specific case studies, such as the introduction of Bt cotton in India or herbicide-resistant soybeans in South America, provides valuable insights into the environmental impacts of GMOs. These examples highlight both the benefits and challenges associated with GMO adoption, informing future decisions and policies.

Ethical and Societal Implications

Ethical Debates Surrounding GMOs

The ethical considerations of GMOs encompass a range of issues, from the manipulation of living organisms to concerns about corporate control over the food supply. These debates often reflect broader societal values and priorities, requiring inclusive dialogue and ethical frameworks to guide decision-making.

Public Perception and Misinformation

Public perception of GMOs varies widely, influenced by cultural, social, and economic factors. Misinformation and lack of understanding can lead to resistance and fear. Effective education and transparent communication are essential to address these concerns and build public trust in GMO technology.

Socioeconomic Impacts on Small-Scale Farmers

The adoption of GMOs can have significant socioeconomic impacts, particularly on small-scale farmers. While GMOs can increase productivity and reduce costs, there are concerns about access, equity, and dependence on seed companies. Policies that support smallholder farmers and ensure fair distribution of benefits are crucial.

Food Sovereignty and GMOs

Food sovereignty, the right of people to define their own food systems, is a key consideration in the GMO debate. Ensuring that communities have control over their agricultural practices and choices is essential for respecting cultural diversity and promoting sustainable development.

Consumer Choice and Labeling

Transparency and informed consumer choice are fundamental principles in the GMO debate. Labeling GMOs allows consumers to make decisions based on their values and preferences. Effective labeling policies should be based on clear, science-based criteria and supported by public education.

Regulatory Frameworks and Safety Assessments

Overview of Global Regulatory Approaches

Regulatory approaches to GMOs vary widely across the world. In the USA, the regulatory framework focuses on the end product rather than the process, while the EU has stricter regulations based on the precautionary principle. Understanding these differences is essential for navigating international markets and trade.

Safety Testing and Risk Assessment Protocols

Robust safety testing and risk assessment protocols are critical for ensuring the safety of GMOs. These protocols include environmental impact assessments, allergenicity testing, and long-term health studies. Innovations in regulatory science, such as genome editing-specific regulations, are needed to keep pace with technological advancements.

Innovations in Regulatory Science

As genetic engineering techniques evolve, regulatory frameworks must adapt to new challenges and opportunities. Precision breeding regulations, which focus on the specific modifications rather than the technology used, offer a promising approach. Harmonizing these standards globally will facilitate innovation while ensuring safety.

Challenges in Harmonizing International Standards

Differences in regulatory standards pose challenges for the global trade of GMOs. Efforts to harmonize these standards must balance the need for innovation with public safety and ethical considerations. International collaboration and dialogue are essential for achieving these goals.

Case Studies of Regulatory Successes and Failures

Examining case studies of regulatory successes and failures provides valuable lessons for future policy development. Examples such as the approval of Golden Rice in the Philippines or the controversy over GMO salmon in North America highlight the complexities of regulating GMOs.

Public Perception and Communication Strategies

Understanding Public Concerns and Misconceptions

Addressing public concerns and misconceptions about GMOs requires a nuanced understanding of the underlying issues. Factors such as cultural values, risk perception, and trust in institutions play a significant role. Engaging with these concerns through dialogue and education is essential for building public trust.

Role of Media and Social Networks in Shaping Public Opinion

The media and social networks have a profound impact on public opinion about GMOs. Misinformation and sensationalism can exacerbate fears and resistance. Scientists and policymakers must leverage these platforms to provide accurate information and counteract myths.

Effective Communication Strategies for Scientists and Policymakers

Effective communication strategies involve transparency, honesty, and empathy. Scientists and policymakers should engage with the public in an open and respectful manner, addressing concerns and providing evidence-based information. Building relationships with trusted community leaders can also enhance credibility.

Case Studies of Successful Public Engagement Initiatives

Examples of successful public engagement initiatives, such as community-based participatory research or public forums on GMOs, provide valuable insights into effective strategies. These case studies highlight the importance of inclusive dialogue and collaboration in building public trust and acceptance.

The Role of GMOs in Sustainable Agriculture and Food Security

GMOs and the UN Sustainable Development Goals

GMOs have the potential to contribute to several UN Sustainable Development Goals (SDGs), including zero hunger, good health and well-being, and climate action. Integrating GMOs into broader sustainability strategies can enhance their impact and address global challenges.

Potential of GMOs in Addressing Global Hunger and Malnutrition

GMOs can play a significant role in addressing global hunger and malnutrition by increasing crop yields, enhancing nutritional content, and reducing losses from pests and diseases. Targeted interventions and equitable access are essential for maximizing these benefits.

Integrating GMOs into Agroecological Practices

Agroecology, which emphasizes ecological principles and sustainable farming practices, can benefit from the integration of GMOs. Combining genetic engineering with agroecological approaches can enhance resilience, reduce chemical inputs, and promote biodiversity.

Long-Term Sustainability and GMO Adoption

Ensuring the long-term sustainability of GMO adoption requires a holistic approach that considers environmental, social, and economic factors. Policies and practices should promote sustainable farming systems, support smallholder farmers, and address ethical considerations.

Future Prospects and Challenges

Emerging Trends and Research Directions

Emerging trends in GMO research include the development of gene-edited crops with enhanced traits, synthetic biology applications, and the use of GMOs for environmental remediation. Keeping abreast of these trends is essential for anticipating future developments and opportunities.

Balancing Innovation with Precaution

Balancing innovation with precaution is a key challenge in the future of GMOs. While the potential benefits are significant, careful consideration of risks and ethical implications is essential. Adopting a precautionary approach that allows for responsible innovation is crucial.

The Role of International Collaboration

International collaboration is vital for advancing GMO research, harmonizing regulatory standards, and addressing global challenges. Collaborative efforts can enhance the exchange of knowledge, resources, and best practices, promoting a more sustainable and equitable future.

Potential Barriers to the Adoption of GMOs

Barriers to the adoption of GMOs include regulatory hurdles, public resistance, and socioeconomic challenges. Addressing these barriers requires coordinated efforts from scientists, policymakers, industry, and civil society.

Vision for the Future: Scenarios and Projections

The future of GMOs will be shaped by a range of scenarios, from widespread adoption and integration into sustainable agriculture to ongoing debates and resistance. Projections should consider technological advancements, regulatory developments, and societal trends to anticipate future outcomes.

Conclusion

The future of genetically modified organisms is both promising and complex, offering significant potential benefits alongside notable challenges. Technological advancements are opening new possibilities for enhancing food security, environmental sustainability, and public health. However, ethical, ecological, and societal considerations must be carefully addressed to ensure responsible and equitable use of GMOs. By fostering transparent communication, robust regulatory frameworks, and inclusive dialogue, we can navigate the future of GMOs in a way that maximizes their benefits while minimizing risks, contributing to a more sustainable and secure global food system.

References

Blaustein, S. (2007). Splitting Genes: the Future of Genetically Modified Organisms in the Wake of the WTO/Cartagena Standoff. Penn St. Envtl. L. Rev., 16, 367.

Gupta, R., & Singh, R. L. (2017). Genetically modified organisms (GMOs) and environment. Principles and Applications of Environmental Biotechnology for a Sustainable Future, 425-465.

Holst-Jensen, A., Rønning, S. B., Løvseth, A., & Berdal, K. G. (2003). PCR technology for screening and quantification of genetically modified organisms (GMOs). Analytical and Bioanalytical Chemistry, 375, 985-993.

Jurkiewicz, A., Zagórski, J., Bujak, F., Lachowski, S., & Florek-Luszczki, M. (2014). Emotional attitudes of young people completing secondary schools towards genetic modification of organisms (GMO) and genetically modified foods (GMF). Annals of agricultural and environmental medicine, 21(1).

Rzymski, P., & Królczyk, A. (2016). Attitudes toward genetically modified organisms in Poland: to GMO or not to GMO?. Food Security, 8, 689-697.

Saxena, G., Kishor, R., Saratale, G. D., & Bharagava, R. N. (2020). Genetically modified organisms (GMOs) and their potential in environmental management: constraints, prospects, and challenges. Bioremediation of Industrial Waste for Environmental Safety: Volume II: Biological Agents and Methods for Industrial Waste Management, 1-19.

Waigmann, E., Paoletti, C., Davies, H., Perry, J., Kärenlampi, S., & Kuiper, H. (2012). Risk assessment of genetically modified organisms (GMOs). EFSA Journal, 10(10), s1008.

19

Renewable Energy in Agriculture Powering the Future

Abstract

Renewable energy in agriculture represents a transformative shift toward sustainability and efficiency in food production systems. This article explores the various renewable energy technologies that are being integrated into agricultural practices, including solar, wind, biomass, and geothermal energy. It examines the benefits of these technologies in reducing greenhouse gas emissions, lowering energy costs, and enhancing the resilience of agricultural operations against climate change. Additionally, the article discusses the challenges and barriers to widespread adoption, such as initial investment costs, technological limitations, and policy obstacles. Through case studies and real-world examples, the article highlights successful implementations of renewable energy in agriculture and offers insights into future trends and innovations. The integration of renewable energy into agriculture not only supports environmental sustainability but also promotes economic viability and energy independence for farmers, paving the way for a more resilient and sustainable agricultural sector.

Introduction

Renewable energy in agriculture refers to the use of sustainable energy sources such as solar, wind, biomass, geothermal, and hydro energy to power agricultural operations. This integration represents a critical step toward achieving environmental sustainability, reducing dependency on fossil fuels, and enhancing the economic viability of farming practices. As the global energy landscape evolves, the agricultural sector faces mounting pressure to adopt cleaner energy solutions to mitigate climate change and ensure food security. This article aims to provide an in-depth analysis of the current state and future potential of renewable energy in agriculture, highlighting the benefits, challenges, and real-world applications of various renewable energy technologies.

Types of Renewable Energy in Agriculture

Solar Energy

Solar energy is one of the most widely adopted renewable energy sources in agriculture due to its versatility and scalability. Photovoltaic (PV) systems convert sunlight directly into electricity, which can be used to power irrigation systems, greenhouses, and other farm operations. Solar thermal applications, such as solar water heaters, provide cost-effective solutions for heating water and buildings on farms.

Wind Energy

Wind energy is harnessed through the use of wind turbines, which convert kinetic energy from the wind into electrical power. Small-scale wind turbines are suitable for individual farms, while large-scale wind farms can supply energy to multiple agricultural operations. Wind energy is particularly effective in regions with consistent wind patterns, providing a reliable source of clean energy.

Biomass Energy

Biomass energy involves the conversion of organic materials, such as crop residues, animal manure, and dedicated energy crops, into energy. Biogas production through anaerobic digestion is a common method, generating methane that can be used for heating, electricity, and as a vehicle fuel. Biofuel crops, such as switchgrass and miscanthus, can be cultivated specifically for energy production, providing a renewable source of biomass.

Geothermal Energy

Geothermal energy utilizes the Earth's natural heat to generate power or provide direct heating applications. In agriculture, geothermal heat pumps can be used to regulate temperatures in greenhouses, livestock buildings, and soil heating systems. Direct use applications, such as geothermal hot water systems, offer sustainable heating solutions for various agricultural needs.

Hydro Energy

Hydro energy harnesses the power of flowing water to generate electricity. Micro-hydro systems are suitable for small-scale agricultural operations, particularly in regions with accessible water sources. These systems can be integrated with irrigation channels to provide a dual-purpose solution, generating energy while delivering water to crops.

Benefits of Renewable Energy in Agriculture

Environmental Benefits

Renewable energy significantly reduces greenhouse gas emissions by replacing fossil fuel-based energy sources. This transition helps mitigate climate change and reduces the environmental footprint of agricultural operations. Additionally, renewable energy technologies often utilize local resources, promoting the conservation of natural habitats and biodiversity.

Economic Benefits

Adopting renewable energy can lead to substantial cost savings for farmers through reduced energy bills and increased energy efficiency. Financial incentives, such as tax credits, grants, and subsidies, further enhance the economic viability of renewable energy projects. Moreover, renewable energy enhances energy security and independence, protecting farmers from volatile energy prices and supply disruptions.

Operational Benefits

Renewable energy technologies enhance the resilience of agricultural operations by providing reliable and sustainable energy sources. This resilience is particularly important in the face of climate change, which poses significant risks to agricultural productivity. Additionally, renewable energy can improve farm productivity and efficiency by powering advanced agricultural technologies and reducing operational costs.

Challenges and Barriers to Adoption

Financial Barriers

One of the primary challenges to the adoption of renewable energy in agriculture is the high initial investment costs associated with installing renewable energy systems. Access to financing and subsidies can be limited, particularly for small-scale farmers. Developing innovative financing models and increasing the availability of financial support are crucial for overcoming these barriers.

Technological Barriers

Integrating renewable energy technologies with existing agricultural systems can be technically challenging. Issues such as energy storage, grid integration, and technological reliability must be addressed to ensure the effective deployment of renewable energy. Continuous research and development are needed to improve the performance and reliability of renewable energy technologies.

Policy and Regulatory Barriers

Inconsistent policies and regulations can hinder the adoption of renewable energy in agriculture. A lack of supportive infrastructure, such as grid connectivity and energy storage facilities, further complicates the implementation of renewable energy projects. Clear and consistent policies, along with supportive regulatory frameworks, are essential for promoting the widespread adoption of renewable energy.

Social and Cultural Barriers

Resistance to change and a lack of awareness and education about renewable energy technologies can impede their adoption in agriculture. Engaging with farmers and stakeholders through educational programs and demonstration projects can help build acceptance and understanding of the benefits of renewable energy.

Case Studies and Real-World Examples

Solar-Powered Farms in California

California is a leading example of the successful integration of solar energy in agriculture. Many farms in the state have adopted solar PV systems to power irrigation pumps, greenhouses, and processing facilities. These installations have led to significant cost savings and reduced carbon emissions, demonstrating the potential of solar energy to transform agricultural operations.

Wind Energy in European Agriculture

Europe has made substantial investments in wind energy, with many agricultural operations benefiting from this clean energy source. Small-scale wind turbines on farms provide electricity for various applications, while larger wind farms contribute to the overall energy grid. The success of wind energy in Europe showcases the viability of this technology for sustainable agriculture.

Biomass and Biogas in Indian Agriculture

India has embraced biomass and biogas technologies to address energy needs in rural and agricultural areas. Biogas plants using animal manure and crop residues generate methane for cooking, heating, and electricity. These projects have improved energy access and reduced environmental pollution, highlighting the benefits of biomass energy for sustainable rural development.

Geothermal Greenhouses in Iceland

Iceland's geothermal resources have enabled the development of geothermal greenhouses, which use the Earth's natural heat to maintain optimal growing conditions year-round. This approach has extended the growing season and increased the variety of crops that can be cultivated, demonstrating the potential of geothermal energy for enhancing agricultural productivity.

Hydro Energy in Small-Scale Irrigation Systems in Africa

Micro-hydro systems integrated with irrigation channels in African countries have provided dual benefits of energy generation and water delivery for crops. These systems have improved water management, increased agricultural yields, and provided a reliable source of electricity for rural communities, showcasing the potential of hydro energy for sustainable development.

Future Trends and Innovations

Advances in Renewable Energy Technologies

Technological advancements, such as smart grids and energy storage solutions, are expected to enhance the efficiency and reliability of renewable energy systems. Precision agriculture technologies, which use data and automation to optimize farming practices, can be powered by renewable energy, further improving agricultural productivity and sustainability.

Policy and Market Trends

Global policy initiatives, such as the Paris Agreement and the UN Sustainable Development Goals, are driving the adoption of renewable energy. Emerging markets, particularly in developing countries, present significant opportunities for investment in renewable energy projects. Supportive policies and market mechanisms will be essential for scaling up renewable energy adoption in agriculture.

Integration with Sustainable Agricultural Practices

Integrating renewable energy with sustainable agricultural practices, such as organic farming and agroecology, can create synergies that enhance environmental and economic benefits. Renewable energy can support sustainable farming practices by providing clean energy for low-impact agricultural technologies and reducing reliance on chemical inputs.

Conclusion

Renewable energy has the potential to transform agriculture by providing sustainable, reliable, and cost-effective energy solutions. The integration of renewable energy technologies can reduce greenhouse gas emissions, lower energy costs, and enhance the resilience of agricultural operations against climate change. However, realizing this potential requires overcoming financial, technological, policy, and social barriers. By fostering innovation, supportive policies, and stakeholder engagement, we can pave the way for a more resilient and sustainable agricultural sector. The future of agriculture lies in harnessing the power of renewable energy to build a sustainable and secure food system for generations to come.

References

Chel, A., & Kaushik, G. (2011). Renewable energy for sustainable agriculture. Agronomy for sustainable development, 31, 91-118.

Fischer, J. R., Finnell, J. A., & Lavoie, B. D. (2006). Renewable energy in Agriculture: Back to the future?. Choices, 21(1), 27-31.

Majeed, Y., Khan, M. U., Waseem, M., Zahid, U., Mahmood, F., Majeed, F., ... & Raza, A. (2023). Renewable energy as an alternative source for energy management in agriculture. Energy Reports, 10, 344-359.

Moriarty, P., & Honnery, D. (2016). Can renewable energy power the future?. Energy policy, 93, 3-7.

Ragazzoni, A., Castellini, A., & Pirazzoli, C. (2014). Production of renewable energy in agriculture: the current situation and future developments. Politica Agricola Internazionale-International Agricultural Policy, 2013(4).

Rahman, M. M., Khan, I., Field, D. L., Techato, K., & Alameh, K. (2022). Powering agriculture: Present status, future potential, and challenges of renewable energy applications. Renewable Energy, 188, 731-749.

Skeer, J., & Leme, R. (2019). Renewable energy in the energy future. In Agriculture & Food Systems to 2050: Global Trends, Challenges and Opportunities (pp. 473-502).

20

Digital Transformation The Rise of Smart Farming

Abstract

The agricultural sector is undergoing a significant transformation, driven by the advent of digital technologies and smart farming practices. This article explores the concept of smart farming, which leverages advanced technologies such as the Internet of Things (IoT), artificial intelligence (AI), big data analytics, and robotics to enhance agricultural productivity, sustainability, and efficiency. It delves into the various components of smart farming, including precision agriculture, smart irrigation systems, and automated machinery. The article also examines the benefits and challenges associated with the digital transformation of agriculture, including economic impacts, environmental considerations, and the role of policy and regulation. Through case studies and real-world examples, the article highlights how smart farming is reshaping the agricultural landscape and paving the way for a more sustainable and food-secure future.

Introduction

The agricultural sector is on the brink of a revolutionary transformation, spurred by the integration of digital technologies into traditional farming practices. Smart farming, also known as precision agriculture or digital agriculture, represents the confluence of advanced technologies and modern agricultural techniques aimed at enhancing productivity, sustainability, and efficiency. This article explores the various dimensions of smart farming, highlighting the technologies involved, the benefits and challenges, and the potential impact on global food security and environmental sustainability. By examining real-world examples and future trends, the article provides a comprehensive overview of the rise of smart farming and its implications for the future of agriculture.

Components of Smart Farming

Precision Agriculture

Precision agriculture is a key component of smart farming, focusing on the precise application of inputs such as water, fertilizers, and pesticides based on real-time data and site-specific conditions. Technologies such as GPS, sensors, and drones enable farmers to monitor and manage their fields with unprecedented accuracy. Applications of precision agriculture include variable rate technology (VRT), which allows for the tailored application of inputs, and remote sensing, which provides detailed information on crop health and soil conditions. The benefits of precision agriculture are manifold, including increased yields, reduced input costs, and minimized environmental impact.

IoT and Smart Sensors

The Internet of Things (IoT) plays a crucial role in smart farming by connecting various devices and sensors to collect and exchange data. Smart sensors, including soil moisture sensors, weather stations, and crop health monitors, provide real-time information that helps farmers make informed decisions. IoT-enabled systems can automate processes such as irrigation and fertilization, ensuring optimal resource use and improving overall farm efficiency. The integration of IoT in agriculture not only enhances productivity but also enables better monitoring and management of environmental conditions.

Artificial Intelligence and Machine Learning

Artificial intelligence (AI) and machine learning (ML) are transforming agriculture by enabling predictive analytics, crop modeling, and automated decision-making. AI algorithms analyze large datasets to identify patterns and make predictions about crop yields, pest outbreaks, and weather conditions. Machine learning models can detect diseases and pests from images, providing early warning and enabling timely interventions. Decision support systems powered by AI help farmers optimize their practices, from planting to harvesting, enhancing both productivity and sustainability.

Robotics and Automation

The use of robotics and automation in agriculture is revolutionizing farm operations. Automated machinery such as tractors, harvesters, and planters reduce labor requirements and increase efficiency. Drones and unmanned aerial vehicles (UAVs) are used for tasks such as crop monitoring, spraying, and mapping. Robotics also play a significant role in livestock management, with automated systems for feeding, milking, and health monitoring. The adoption of robotics in agriculture not only improves operational efficiency but also addresses labor shortages and enhances precision in farm activities.

Big Data Analytics

Big data analytics is essential for smart farming, enabling the processing and analysis of vast amounts of data generated by various sensors and devices.

Data sources include satellite imagery, weather data, soil samples, and crop performance records. Advanced analytics tools help farmers predict yields, optimize resource use, and make informed decisions. By harnessing the power of big data, farmers can enhance productivity, reduce costs, and improve sustainability. Data-driven insights also support research and development in agriculture, fostering innovation and continuous improvement.

Smart Irrigation Systems

Smart irrigation systems use advanced technologies to optimize water use in agriculture. These systems incorporate sensors, weather forecasts, and automated controllers to deliver the right amount of water at the right time. Technologies such as drip irrigation, automated sprinklers, and soil moisture sensors ensure efficient water use, reducing wastage and improving crop health. Smart irrigation not only enhances water conservation but also increases yields and reduces the environmental impact of farming.

Benefits of Smart Farming

Increased Productivity and Efficiency

Smart farming technologies enable farmers to optimize their practices, leading to increased productivity and efficiency. Precision agriculture techniques ensure that inputs are used effectively, reducing waste and enhancing yields. Automated machinery and robotics streamline farm operations, reducing labor requirements and increasing operational efficiency. The integration of AI and big data analytics provides farmers with actionable insights, helping them make informed decisions and optimize their practices.

Environmental Sustainability

One of the key benefits of smart farming is its potential to enhance environmental sustainability. Precision agriculture and smart irrigation systems reduce the use of water, fertilizers, and pesticides, minimizing the environmental impact of farming. By optimizing resource use, smart farming practices help conserve natural resources and reduce pollution. Additionally, technologies such as drones and remote sensing enable better monitoring and management of environmental conditions, supporting sustainable agricultural practices.

Economic Benefits

Smart farming offers significant economic benefits for farmers and the agricultural sector as a whole. The adoption of advanced technologies leads to cost savings through reduced input use and increased efficiency. Improved productivity and yields enhance profitability, providing farmers with better returns on investment. Smart farming also opens up new market opportunities, enabling farmers to compete more effectively in the global market. The economic benefits of smart farming extend beyond individual farmers, contributing to the overall growth and development of the agricultural sector.

Enhancing Food Security

Smart farming has the potential to enhance food security by increasing food production and reducing post-harvest losses. Precision agriculture techniques improve crop yields, ensuring a stable and abundant food supply. Smart irrigation systems and other technologies help farmers adapt to changing climate conditions, increasing resilience and reducing vulnerability to droughts and other extreme weather events. By improving productivity and reducing losses, smart farming contributes to global food security, addressing the challenges of feeding a growing population.

Challenges and Barriers

Technological Challenges

The integration of various technologies in smart farming poses several challenges. Ensuring interoperability and seamless data exchange between different systems and devices is a key technical hurdle. Data management and analysis require sophisticated tools and expertise, which may not be readily available to all farmers. Additionally, the rapid pace of technological advancements necessitates continuous learning and adaptation, presenting a challenge for farmers and agricultural professionals.

Economic and Financial Barriers

The high initial investment costs associated with smart farming technologies can be a significant barrier to adoption, especially for small-scale and resource-poor farmers. Access to funding and credit is essential to overcome these financial barriers and enable farmers to invest in new technologies. Economic disparities among farmers can exacerbate these challenges, making it difficult for some to benefit from the advancements in smart farming.

Social and Cultural Barriers

Resistance to change and adoption of new technologies can be a significant barrier to the digital transformation of agriculture. Social and cultural factors, including traditional farming practices and lack of awareness, can hinder the adoption of smart farming technologies. Education and training are crucial to address these barriers, providing farmers with the knowledge and skills needed to implement and benefit from smart farming practices. Additionally, the impact of automation and digital technologies on rural communities and labor markets must be carefully managed to ensure equitable and inclusive development.

Policy and Regulatory Challenges

The lack of standardized regulations and supportive policies can impede the adoption of smart farming technologies. Clear and consistent regulatory frameworks are needed to address issues such as data privacy, security, and intellectual property rights. Policies that provide incentives and support for the adoption of smart farming technologies can also play a crucial role in driving digital transformation in agriculture. International collaboration and harmonization of standards are essential to facilitate the global trade and implementation of smart farming practices.

Case Studies and Real-World Examples

Case Study 1: Precision Agriculture in the United States

Precision agriculture has been widely adopted in the United States, leveraging advanced technologies to optimize farm operations. Farmers use GPS-guided machinery, drones, and sensors to monitor and manage their fields with precision. The implementation of variable rate technology and remote sensing has led to significant improvements in productivity and resource efficiency. Case studies from the Midwest, where precision agriculture is prevalent, demonstrate the benefits of increased yields, reduced input costs, and enhanced environmental sustainability.

Case Study 2: Smart Irrigation in Australia

Australia has implemented smart irrigation systems to address water scarcity and improve agricultural productivity. Technologies such as automated sprinklers, soil moisture sensors, and weather forecasting tools are used to optimize water use. The adoption of smart irrigation has resulted in significant water savings, improved crop health, and increased yields. Case studies from regions such as the Murray-Darling Basin highlight the positive impact of smart irrigation on water conservation and agricultural sustainability.

Case Study 3: Robotics and Automation in Japan

Japan has been a leader in adopting robotics and automation in agriculture, driven by labor shortages and the need for increased efficiency. Automated machinery, such as robotic harvesters and planters, are used to perform labor-intensive tasks. Drones and UAVs are deployed for crop monitoring and spraying. The implementation of robotics in agriculture has improved operational efficiency, reduced labor requirements, and enhanced productivity.

Case studies from Japanese farms illustrate the economic and labor market impacts of robotics and automation.

Case Study 4: IoT and Data Analytics in the Netherlands

The Netherlands has embraced IoT and data analytics in agriculture, leveraging advanced technologies to enhance productivity and sustainability. Farmers use IoT-enabled sensors to monitor soil conditions, weather, and crop health in real-time. Data collected from these sensors is analyzed to provide actionable insights and support decision-making. The integration of IoT and data analytics has led to improved resource use, higher yields, and better environmental management. Case studies from Dutch farms showcase the benefits of data-driven decision-making in agriculture.

Future Trends and Innovations

Emerging Technologies in Smart Farming

Emerging technologies such as blockchain, advanced AI, and next-generation sensors are poised to further transform smart farming. Blockchain technology can enhance supply chain transparency, ensuring the traceability and authenticity of agricultural products. Advanced AI and machine learning applications will enable more precise and predictive analytics, enhancing decision-making and optimization. Next-generation sensors and IoT devices will provide more accurate and comprehensive data, supporting real-time monitoring and management of agricultural operations.

The Role of Startups and Innovation Hubs

Agri-tech startups and innovation hubs are playing a crucial role in driving the digital transformation of agriculture. Startups are developing innovative solutions, from AI-powered crop monitoring systems to robotic harvesters. Innovation hubs and research centers provide a platform for collaboration, knowledge sharing, and the development of new technologies. The role of startups and innovation hubs in fostering innovation and driving the adoption of smart farming practices is essential for the future of agriculture.

Collaborative Efforts and Partnerships

Collaborative efforts and partnerships are vital for advancing smart farming technologies and practices. Public-private partnerships, international collaborations, and multi-stakeholder initiatives can enhance the exchange of knowledge, resources, and best practices. Collaborative efforts can also address the challenges of scaling and implementing smart farming technologies, ensuring that the benefits are widely distributed and accessible to all farmers.

Vision for the Future of Smart Farming

The future of smart farming is characterized by a vision of sustainable, efficient, and resilient agriculture. Technological advancements will continue to drive innovation, enhancing productivity and environmental sustainability. Collaborative efforts and supportive policies will enable the widespread adoption of smart farming practices, addressing the challenges of food security and climate change. The vision for the future of smart farming is one of inclusive and equitable development, where digital technologies empower farmers and contribute to a sustainable and food-secure world.

Conclusion

The digital transformation of agriculture through smart farming represents a significant shift towards more sustainable, efficient, and productive farming practices. The integration of advanced technologies such as IoT, AI, robotics, and big data analytics is reshaping the agricultural landscape, offering numerous benefits while also presenting challenges that need to be addressed. By leveraging the potential of smart farming, we can enhance food security, promote environmental sustainability, and support the economic development of the agricultural sector. The future of smart farming depends on the collaborative efforts of all stakeholders, including farmers, policymakers, researchers, and technology providers, to create a more sustainable and food-secure world.

References

Dayıoğlu, M. A., & Turker, U. (2021). Digital transformation for sustainable future-agriculture 4.0: A review. Journal of Agricultural Sciences, 27(4), 373-399.

Holzinger, A., Saranti, A., Angerschmid, A., Retzlaff, C. O., Gronauer, A., Pejakovic, V., & Stampfer, K. (2022). Digital transformation in smart farm and forest operations needs human-centered AI: challenges and future directions. Sensors, 22(8), 3043.

Massruhá, S. M. F. S., Leite, M. D. A., Luchiari Junior, A., & Evangelista, S. R. M. (2023). Digital transformation in the field towards sustainable and smart agriculture.

Rao, V. P., & Anitha, V. (2021). Agriculture transformation through Big Data and smart farming. International Journal of Innovative Horticulture, 10(2), 130-137.

Rimma, Z., Marina, T., Olga, R., Andrey, C., & Albina, S. (2020, March). Major Trends in the Digital Transformation of Agriculture. In "New Silk Road: Business Cooperation and Prospective of Economic Development"(NSRBCPED 2019) (pp. 271-275). Atlantis Press.

Rodríguez, M. A., Cuenca, L., & Ortiz, Á. (2019). Big data transformation in agriculture: From precision agriculture towards smart farming. In Collaborative Networks and Digital Transformation: 20th IFIP WG 5.5 Working Conference on Virtual Enterprises, PRO-VE 2019, Turin, Italy, September 23–25, 2019, Proceedings 20 (pp. 457-474). Springer International Publishing.

21

Food Security Challenges and Solutions

Abstract

Food security, defined as the availability, access, and utilization of sufficient, safe, and nutritious food for all people at all times, is a critical global challenge. This article examines the multifaceted issues contributing to food insecurity, including population growth, climate change, economic instability, and geopolitical conflicts. It explores the impact of these factors on food production, distribution, and access, highlighting the interconnectedness of global systems. Furthermore, the article discusses various solutions and strategies to enhance food security, such as sustainable agricultural practices, technological innovations, policy reforms, and international cooperation. Emphasizing a holistic and integrated approach, the article argues that addressing food security requires coordinated efforts across sectors and scales, from local communities to global networks. By fostering resilience, equity, and sustainability in food systems, we can move towards a future where everyone has access to the nourishment they need to lead healthy and productive lives.

Introduction

Food security is defined by the Food and Agriculture Organization (FAO) as a situation where all people, at all times, have physical, social, and economic access to sufficient, safe, and nutritious food that meets their dietary needs and food preferences for an active and healthy life. The concept encompasses four dimensions:

1. **Availability**: Sufficient quantities of food of appropriate quality, supplied through domestic production or imports.
2. **Access**: Adequate resources to acquire appropriate foods for a nutritious diet.
3. **Utilization**: Proper biological use of food, requiring a diet that meets nutritional needs, safe food preparation, and adequate health and sanitation services.

4. **Stability**: Having access to adequate food at all times, without risk of losing access due to sudden shocks (economic or climatic crisis) or cyclical events (seasonal food scarcity).

Challenges to Food Security

1. **Population Growth**

   The global population is projected to reach 9.7 billion by 2050, according to the United Nations. This rapid growth increases the demand for food, straining agricultural systems and natural resources. Meeting this demand requires not only increasing food production but also ensuring that food distribution systems are efficient and equitable.

2. **Climate Change**

   Climate change poses a significant threat to food security through its impact on agriculture. Changes in temperature, precipitation patterns, and the increased frequency of extreme weather events (such as droughts, floods, and hurricanes) disrupt food production. For example, droughts can devastate crop yields, while floods can destroy crops and infrastructure, leading to food shortages and increased prices.

3. **Soil Degradation**

   Soil degradation, including erosion, nutrient depletion, and salinization, reduces the land's productivity and its ability to support agriculture. Unsustainable farming practices, deforestation, and industrial activities contribute to soil degradation, compromising future food production and security.

4. **Water Scarcity**

   Agriculture accounts for approximately 70% of global freshwater use. As water resources become increasingly scarce due to over-extraction, pollution, and climate change, the availability of water for irrigation and food production is threatened. Efficient water management practices are essential to maintain food production in water-scarce regions.

5. **Economic and Political Instability**

   Economic and political instability can disrupt food production and distribution systems, leading to food shortages and price volatility. Conflicts, trade restrictions, and economic downturns affect farmers' ability to produce and sell their products, while also impacting consumers' purchasing power and access to food.

6. **Inefficient Food Systems**

   Inefficiencies in food production, processing, and distribution contribute to food loss and waste. According to the FAO, approximately one-third of all food produced globally is lost or wasted. Improving efficiency across the food supply chain is critical to ensuring that more food reaches consumers.

7. **Inequality and Access to Food**

   Social and economic inequalities affect people's ability to access nutritious food. Marginalized populations, including low-income households, rural communities, and ethnic minorities, often face higher rates of food insecurity. Addressing these disparities is crucial to achieving food security for all.

Solutions to Food Security

1. **Sustainable Agriculture**

   Promoting sustainable agricultural practices is essential to increase food production while preserving natural resources. Sustainable practices include:

   - **Agroecology**: Integrating ecological principles into agricultural systems to enhance biodiversity, improve soil health, and reduce the need for chemical inputs.
   - **Conservation Agriculture**: Practices such as minimal soil disturbance, crop rotation, and cover cropping to maintain soil structure and fertility.
   - **Organic Farming**: Avoiding synthetic fertilizers and pesticides, using organic inputs, and fostering ecological balance.
   - **Agroforestry**: Combining agriculture with tree planting to improve soil health, increase biodiversity, and provide additional sources of income.

2. **Climate-Resilient Crops and Practices**

   Developing and promoting climate-resilient crops and farming practices can help mitigate the impacts of climate change on agriculture. These include:

   - **Drought-Resistant Crops**: Breeding and genetically modifying crops to withstand periods of low water availability.
   - **Flood-Tolerant Varieties**: Developing crops that can survive and produce yields under waterlogged conditions.

- **Climate-Smart Agriculture**: Practices that increase productivity, enhance resilience, and reduce greenhouse gas emissions, such as improved irrigation techniques, precision farming, and integrated pest management.

3. **Efficient Water Management**

   Improving water use efficiency in agriculture is critical to addressing water scarcity. Solutions include:

   - **Drip Irrigation**: Delivering water directly to the plant roots, reducing water wastage and evaporation.
   - **Rainwater Harvesting**: Collecting and storing rainwater for agricultural use during dry periods.
   - **Water Recycling**: Reusing treated wastewater for irrigation.
   - **Improved Irrigation Practices**: Adopting techniques such as deficit irrigation and scheduling to optimize water use.

4. **Enhancing Food Systems Efficiency**

   Reducing food loss and waste throughout the food supply chain can significantly improve food security. Strategies include:

   - **Post-Harvest Technologies**: Developing and implementing technologies for better storage, transportation, and processing to reduce losses.
   - **Supply Chain Optimization**: Enhancing logistics and infrastructure to ensure efficient distribution and reduce spoilage.
   - **Consumer Education**: Raising awareness about food waste and encouraging responsible consumption practices.

5. **Economic and Social Policies**

   Implementing policies that address economic and social inequalities can improve access to food. These include:

   - **Social Safety Nets**: Programs such as food subsidies, cash transfers, and food assistance to support vulnerable populations.
   - **Inclusive Economic Growth**: Promoting policies that create job opportunities and improve incomes for marginalized groups.
   - **Land Reform**: Ensuring equitable access to land and resources for smallholder farmers and marginalized communities.

6. **Strengthening Global Cooperation**

   International cooperation is essential to address the global nature of food security challenges. Collaborative efforts include:

- **Trade Policies**: Promoting fair and open trade to ensure the availability of food in regions with production deficits.
- **Research and Development**: Investing in agricultural research and technology transfer to improve productivity and resilience.
- **Disaster Response**: Coordinating international aid and support during food crises caused by natural disasters or conflicts.

7. **Empowering Women in Agriculture**

   Women play a crucial role in agriculture, particularly in developing countries. Empowering women through access to resources, education, and decision-making can enhance food security. Initiatives include:

   - **Access to Land and Credit**: Ensuring women have equal rights to land ownership and access to financial services.
   - **Education and Training**: Providing women with agricultural education and skills development.
   - **Participation in Decision-Making**: Involving women in agricultural planning and policy-making processes.

8. **Technological Innovations**

   Harnessing technological advancements can revolutionize agriculture and food systems. Innovations include:

   - **Precision Agriculture**: Using sensors, drones, and satellite imagery to optimize farming practices and resource use.
   - **Biotechnology**: Developing genetically modified crops with improved yield, nutritional content, and resistance to pests and diseases.
   - **Digital Platforms**: Connecting farmers to markets, information, and financial services through mobile and internet technologies.

Conclusion

Achieving food security in the face of growing challenges requires a multifaceted approach that integrates sustainable agricultural practices, climate resilience, efficient resource management, economic and social policies, global cooperation, gender empowerment, and technological innovation. By addressing the root causes of food insecurity and implementing holistic solutions, it is possible to ensure that all people have access to sufficient, safe, and nutritious food for a healthy and active life. The path to food security is complex and demanding, but with coordinated efforts and innovative strategies, it is an attainable goal.

References

Chitondo, L., Chanda, C. T. and Phiri, E. V. (2024). Ensuring National Food Security in Southern Africa: Challenges and Solutions. International Journal of Novel Research in Interdisciplinary Studies, 11(2), 1-18.

Chouhan, G. K., Verma, J. P., Jaiswal, D. K., Mukherjee, A., Singh, S., de Araujo Pereira, A. P., and Singh, B. K. (2021). Phytomicrobiome for promoting sustainable agriculture and food security: opportunities, challenges, and solutions. Microbiological Research, 248, 126763.

Hertel, T. W., Baldos, U. L. and Fuglie, K. O. (2020). Trade in technology: A potential solution to the food security challenges of the 21st century. European Economic Review, 127, 103479.

Jennings, S., Stentiford, G. D., Leocadio, A. M., Jeffery, K. R., Metcalfe, J. D., Katsiadaki, I., and Verner-Jeffreys, D. W. (2016). Aquatic food security: insights into challenges and solutions from an analysis of interactions between fisheries, aquaculture, food safety, human health, fish and human welfare, economy and environment. Fish and Fisheries, 17(4), 893-938.

Mbow, C., Van Noordwijk, M., Luedeling, E., Neufeldt, H., Minang, P. A. and Kowero, G. (2014). Agroforestry solutions to address food security and climate change challenges in Africa. Current Opinion in Environmental Sustainability, 6, 61-67.

Mc Carthy, U., Uysal, I., Badia-Melis, R., Mercier, S., O'Donnell, C. and Ktenioudaki, A. (2018). Global food security–Issues, challenges and technological solutions. Trends in Food Science & Technology, 77: 11-20.

22

Women in Agriculture Empowering Female Farmers

Abstract

Women play a crucial role in agriculture worldwide, contributing significantly to food security, economic stability, and community development. Despite their essential role, female farmers face numerous challenges, including limited access to land, credit, education, and technology. This article explores the multifaceted contributions of women in agriculture, the barriers they face, and the strategies for empowering female farmers. By addressing these challenges through targeted interventions and policies, we can enhance agricultural productivity, promote gender equality, and achieve sustainable development goals.

Introduction

Women are indispensable to the agricultural sector, particularly in developing countries where they make up a substantial portion of the agricultural labor force. Their roles range from fieldwork and livestock management to food processing and marketing. However, female farmers often encounter systemic barriers that hinder their productivity and limit their potential. Empowering women in agriculture is not only a matter of social justice but also a critical factor in boosting agricultural output, improving food security, and fostering sustainable development.

The Role of Women in Agriculture

1. **Contribution to Food Production**

   Women are involved in various stages of agricultural production, including planting, weeding, harvesting, and processing. In many regions, they are responsible for growing staple crops that form the backbone of local diets. For instance, in sub-Saharan Africa, women contribute to 60-80% of food production.

2. **Livestock Management**

   Women often manage livestock, which provides essential sources of food, income, and fertilizer for crops. Their involvement in animal husbandry includes feeding, breeding, milking, and healthcare of animals.

3. **Household Food Security**

   Women play a pivotal role in ensuring household food security by managing small-scale farms and gardens. They grow vegetables, fruits, and other crops that contribute to the family's nutritional needs.

4. **Post-Harvest Activities**

   Women are heavily involved in post-harvest activities such as processing, storage, and marketing of agricultural products. These activities are crucial for adding value to produce and ensuring its availability throughout the year.

Challenges Faced by Female Farmers

1. **Limited Access to Land**

   Land ownership and tenure security are significant challenges for women in agriculture. In many cultures, customary laws and practices restrict women's rights to own, inherit, and control land. Without secure land tenure, women are less likely to invest in land improvements and more likely to be excluded from agricultural decision-making processes.

2. **Lack of Access to Credit and Financial Services**

   Access to credit is essential for purchasing inputs like seeds, fertilizers, and equipment. However, female farmers often face discrimination from financial institutions, which view them as high-risk borrowers due to their lack of collateral and formal employment.

3. **Insufficient Agricultural Education and Training**

   Educational disparities limit women's access to agricultural training and extension services. This gap in knowledge and skills restricts their ability to adopt modern farming techniques and innovations that could enhance productivity.

4. **Inadequate Access to Technology**

   Technological advancements in agriculture can significantly improve efficiency and yields. However, women often have limited access to technologies such as improved seeds, irrigation systems, and machinery, partly due to financial constraints and gender biases.

5. **Socio-Cultural Barriers**

   Cultural norms and gender roles often dictate that women take on domestic responsibilities, reducing the time and energy they can devote to agricultural activities. Additionally, societal expectations can limit their participation in decision-making and leadership roles within agricultural communities.

Strategies for Empowering Female Farmers

1. **Land Rights and Tenure Security**

   Ensuring women's access to land and secure tenure is fundamental to empowering female farmers. Legal reforms and policy interventions are needed to protect women's land rights and promote equitable land distribution. Community-based approaches that involve local leaders and stakeholders can also help address customary practices that disadvantage women.

2. **Access to Credit and Financial Services**

   Improving access to credit for women can be achieved through microfinance institutions, women's savings groups, and gender-sensitive financial products. Providing financial literacy training and encouraging women to form cooperatives can also enhance their financial inclusion and bargaining power.

3. **Agricultural Education and Training**

   Investing in agricultural education and extension services tailored to women's needs is crucial. Training programs should focus on practical skills, sustainable farming practices, and business management. Facilitating women's participation in agricultural research and innovation can also drive gender-responsive solutions.

4. **Access to Technology**

   Bridging the technological gap requires initiatives that provide affordable and accessible agricultural technologies to women. Programs that facilitate the sharing of knowledge and best practices, along with the provision of equipment and inputs, can significantly improve women's productivity.

5. **Socio-Cultural Change**

   Promoting gender equality in agriculture involves challenging and changing socio-cultural norms that restrict women's roles. Community awareness campaigns, gender-sensitive policies, and the inclusion of men in gender equality initiatives can help shift perceptions and promote more equitable participation.

Case Studies

1. **Land Rights in Rwanda**

   Rwanda has made significant strides in promoting gender equality in land ownership. The country's land tenure reform program includes provisions for joint land registration for married couples and the legal recognition of women's land rights. This reform has increased women's access to land and improved their economic stability.

2. **Microfinance in Bangladesh**

   The Grameen Bank in Bangladesh has been a pioneer in providing microfinance to women. By offering small loans without collateral, the bank has empowered millions of women to start small businesses and invest in agriculture. This model has been replicated in various countries, demonstrating the impact of financial inclusion on women's empowerment.

3. **Agricultural Training in India**

   The Self Employed Women's Association (SEWA) in India provides agricultural training and support services to women farmers. SEWA's programs focus on sustainable farming practices, market linkages, and financial literacy, enabling women to improve their livelihoods and contribute to household food security.

4. **Technology Access in Kenya**

   The Digital Green initiative in Kenya uses video-based training to disseminate agricultural knowledge and technologies to women farmers. This approach has proven effective in reaching women in remote areas and providing them with the information needed to adopt new farming practices.

Policy Recommendations

1. Integrating Gender into Agricultural Policies

   Governments should integrate gender considerations into national agricultural policies and programs. This includes conducting gender assessments, setting targets for female participation, and allocating resources to gender-responsive initiatives.

2. Supporting Women's Cooperatives

   Encouraging the formation and strengthening of women's cooperatives can enhance their bargaining power, improve access to markets, and facilitate the sharing of resources and knowledge. Cooperative models can also provide a platform for women to advocate for their rights and interests.

3. Enhancing Data Collection and Research

   Improving data collection and research on women in agriculture is essential for designing effective policies and interventions. Gender-disaggregated data can provide insights into the specific challenges and opportunities for female farmers and help track progress towards gender equality goals.

4. Promoting Gender Equality in Agricultural Extension

   Agricultural extension services should be gender-sensitive, ensuring that women have equal access to training, information, and resources. Employing more female extension workers and incorporating gender training for all staff can improve outreach and impact.

5. Ensuring Access to Markets

   Facilitating women's access to markets through infrastructure development, market information systems, and support services can enhance their economic opportunities. Policies that promote fair trade practices and address market barriers can also benefit women farmers.

Conclusion

Empowering women in agriculture is essential for achieving food security, reducing poverty, and promoting sustainable development. By addressing the challenges faced by female farmers and implementing targeted strategies, we can unlock the full potential of women in agriculture. This requires a concerted effort from governments, international organizations, civil society, and the private sector to create an enabling environment for gender equality and inclusive growth in the agricultural sector. Through collective action and commitment, we can ensure that women farmers are recognized, supported, and empowered to contribute to a more sustainable and food-secure world.

References

Altenbuchner, C., Vogel, S., & Larcher, M. (2017, September). Effects of organic farming on the empowerment of women: A case study on the perception of female farmers in Odisha, India. In Women's Studies International Forum (Vol. 64, pp. 28-33). Pergamon.

Anderson, C. L., Reynolds, T. W., Biscaye, P., Patwardhan, V., & Schmidt, C. (2021). Economic benefits of empowering women in agriculture: Assumptions and evidence. The Journal of Development Studies, 57(2), 193-208.

Asadullah, M. N., & Kambhampati, U. (2021). Feminization of farming, food security and female empowerment. Global Food Security, 29, 100532.

Hossain, M., & Jaim, W. (2011, January). Empowering women to become farmer entrepreneur. In IFAD Conference on New Directions for Smallholder Agriculture (Vol. 24, p. 25).

Ingutia, R., & Sumelius, J. (2022). Do farmer groups improve the situation of women in agriculture in rural Kenya?. International Food and Agribusiness Management Review, 25(1), 135-156.

Patel, A. (2012). Empowering women in agriculture. Yojana, 56, 19-22.

Rao, S. (2011). Work and empowerment: Women and agriculture in South India. The Journal of Development Studies, 47(2), 294-315.

23

Farmers' Rights and Advocacy

Abstract

Farmers' rights and advocacy are essential for the sustainability and advancement of agricultural communities worldwide. These rights encompass access to land, seeds, resources, and fair market conditions, along with the recognition of farmers' contributions to food security and biodiversity. Advocacy for farmers' rights involves mobilizing stakeholders, influencing policy, and fostering grassroots movements to ensure equitable treatment and support for farmers. This article delves into the historical context, key components, challenges, and effective strategies for advocating farmers' rights, underscoring the importance of a collective approach to empower the agricultural sector.

Introduction

Farmers are the backbone of global food systems, playing a crucial role in feeding populations and maintaining biodiversity. However, they often face numerous challenges, including land tenure insecurity, lack of access to quality seeds, limited financial resources, and unfavorable market conditions. Recognizing and advocating for farmers' rights is vital to address these issues and support sustainable agricultural practices. This article explores the historical development of farmers' rights, their key components, the challenges farmers face, and effective strategies for advocacy, emphasizing the importance of collective action to empower the agricultural sector.

Historical Context of Farmers' Rights

Early Agricultural Societies

In early agricultural societies, farmers' rights were often determined by local customs and traditions. Land ownership and resource use were governed by community rules, which ensured that farming practices were sustainable and equitable. These traditional systems, while varied, generally supported the notion that land and resources were communal assets that should be managed for the benefit of all members of the community.

Colonial and Post-Colonial Periods

The advent of colonialism significantly disrupted traditional agricultural systems. Colonial powers imposed new land ownership structures, often concentrating land in the hands of a few elites and marginalizing smallholder farmers. This shift led to widespread dispossession and the erosion of traditional farming practices. In the post-colonial period, many newly independent nations struggled to rectify these imbalances, leading to ongoing conflicts over land and resources.

Emergence of the Farmers' Rights Movement

The concept of farmers' rights began to gain prominence in the late 20th century, driven by growing awareness of the importance of agricultural biodiversity and the contributions of smallholder farmers to global food security. The International Treaty on Plant Genetic Resources for Food and Agriculture (ITPGRFA), adopted in 2001, marked a significant milestone in recognizing farmers' rights at the international level. The treaty acknowledges the vital role of farmers in conserving and developing plant genetic resources and emphasizes the need to protect their rights to save, use, exchange, and sell farm-saved seed.

Key Components of Farmers' Rights

1. **Land Rights**

   Secure land tenure is fundamental to farmers' rights. It ensures that farmers have legal recognition and protection of their land ownership or use rights, enabling them to invest in and sustainably manage their land. Secure land tenure also provides farmers with the collateral needed to access credit and other financial services.

2. **Seed Rights**

   Farmers' rights to seeds are crucial for maintaining agricultural biodiversity and ensuring food security. These rights include the ability to save, use, exchange, and sell farm-saved seeds. Protecting farmers' seed rights helps preserve traditional knowledge and practices, which are essential for the adaptation of crops to changing environmental conditions.

3. **Access to Resources**

   Farmers need access to a range of resources, including water, fertilizers, tools, and technology, to improve their productivity and livelihoods. Ensuring equitable access to these resources is critical for supporting smallholder farmers and promoting sustainable agricultural practices.

4. **Fair Market Conditions**

   Fair market conditions are essential for farmers to receive a just return for their labor and produce. This includes access to markets, transparent pricing mechanisms, and protection from exploitation by intermediaries. Fair trade practices and support for local markets can help ensure that farmers receive equitable compensation for their products.

5. **Recognition and Support**

   Recognition of farmers' contributions to food security, biodiversity, and sustainable development is vital. This recognition should be accompanied by support from governments, international organizations, and civil society to create enabling environments for farmers to thrive. Support can include policy frameworks, financial assistance, and capacity-building programs.

Challenges Facing Farmers

1. **Land Tenure Insecurity**

   In many parts of the world, farmers face insecure land tenure, which undermines their ability to invest in sustainable agricultural practices. Land grabs, unclear property rights, and weak legal protections exacerbate this issue, leaving farmers vulnerable to displacement and loss of livelihoods.

2. **Limited Access to Quality Seeds**

   Farmers often struggle to access high-quality seeds that are adapted to local conditions and resistant to pests and diseases. Seed monopolies and restrictive intellectual property laws further limit farmers' ability to save, exchange, and sell seeds, reducing their autonomy and resilience.

3. **Financial Constraints**

   Access to credit and financial services is a significant barrier for many farmers, particularly smallholders. Without financial support, farmers cannot invest in essential inputs, technologies, or infrastructure needed to enhance productivity and adapt to climate change.

4. **Market Barriers**

   Smallholder farmers frequently encounter barriers to accessing markets, including lack of infrastructure, information asymmetry, and exploitation by middlemen. These challenges result in low prices for their produce and reduced incomes, hindering their ability to invest in their farms.

5. **Climate Change**

   Climate change poses a severe threat to agriculture, with increasing frequency and intensity of extreme weather events, changing precipitation patterns, and rising temperatures. These changes disrupt agricultural cycles, reduce crop yields, and increase the vulnerability of farming communities.

6. **Social and Gender Inequities**

   Women and marginalized groups often face additional barriers in accessing resources, land, and support services. Gender inequities in agriculture limit women's ability to contribute fully to food production and benefit from agricultural development.

Strategies for Advocating Farmers' Rights

1. **Grassroots Mobilization**

   Grassroots mobilization is a powerful tool for advocating farmers' rights. By organizing at the local level, farmers can collectively voice their concerns, demand their rights, and influence policy decisions. Grassroots movements often involve forming cooperatives, associations, or unions to strengthen their bargaining power and solidarity.

2. **Legal and Policy Advocacy**

   Advocating for legal and policy reforms is essential to protect and promote farmers' rights. This involves engaging with policymakers, drafting and proposing legislation, and participating in public consultations. Effective advocacy requires a thorough understanding of legal frameworks and the ability to articulate the needs and rights of farmers.

3. **Capacity Building**

   Building the capacity of farmers and their organizations is crucial for effective advocacy. Training programs on leadership, negotiation, legal rights, and sustainable farming practices can empower farmers to advocate for their rights and improve their livelihoods. Capacity-building initiatives should also focus on enhancing the organizational skills of farmer groups to strengthen their collective action.

4. **Strategic Alliances and Partnerships**

   Forming strategic alliances and partnerships with civil society organizations, research institutions, and international bodies can amplify advocacy efforts. Collaborative networks can provide technical expertise, resources, and platforms for sharing knowledge and best practices. Partnerships with media organizations can also help raise public awareness and generate support for farmers' rights.

5. **Participatory Research and Documentation**

   Participatory research involving farmers in the documentation of their experiences, challenges, and innovations can provide valuable evidence for advocacy. This approach ensures that farmers' voices are heard and their knowledge is recognized in policy and decision-making processes. Documenting success stories and best practices can also inspire and guide other farming communities.

6. **Public Awareness Campaigns**

   Raising public awareness about farmers' rights and the importance of sustainable agriculture can garner broad support for advocacy efforts. Campaigns can use various media, including social media, radio, television, and print, to reach diverse audiences. Engaging the public through storytelling, events, and educational programs can help build a more informed and supportive society.

Case Studies in Farmers' Rights Advocacy

1. **The Landless Workers' Movement in Brazil (MST)**

   The Movimento dos Trabalhadores Rurais Sem Terra (MST), or Landless Workers' Movement, is one of the largest social movements in Latin America, advocating for land reform and the rights of landless farmers in Brazil. MST has successfully mobilized millions of members to occupy unused land, pressuring the government to redistribute land to landless families. The movement's advocacy has led to the establishment of numerous settlements, improving the livelihoods of thousands of families and promoting sustainable agricultural practices.

2. **Navdanya in India**

   Navdanya, a non-governmental organization founded by Dr. Vandana Shiva, advocates for farmers' rights, biodiversity conservation, and sustainable agriculture in India. The organization works to protect farmers' seed rights through community seed banks, training programs, and legal advocacy against corporate control of seeds. Navdanya's efforts have empowered farmers to reclaim their traditional seed varieties, enhance food security, and promote ecological farming practices.

3. **La Via Campesina**

   La Via Campesina is an international movement of peasants, small-scale farmers, agricultural workers, and indigenous peoples advocating for food sovereignty and farmers' rights. With members from over 80 countries, La Via Campesina campaigns for equitable access to land, seeds, water, and other resources. The movement has played a crucial role in influencing international policies, such as the United Nations Declaration on the Rights of Peasants and Other People Working in Rural Areas (UNDROP).

4. **The African Union's Comprehensive Africa Agriculture Development Programme (CAADP)**

   The Comprehensive Africa Agriculture Development Programme (CAADP) is a continental initiative led by the African Union to improve food security and agricultural development in Africa. CAADP emphasizes the need for inclusive policies that recognize and support the rights of smallholder farmers. Through regional and national compacts, CAADP promotes investments in agriculture, capacity building, and policy reforms to empower farmers and enhance agricultural productivity.

Policy Recommendations

1. **Strengthening Land Rights**

   Governments should implement policies and legal frameworks that secure land tenure for farmers, particularly smallholders and marginalized groups. Land reforms should prioritize equitable distribution, recognition of customary land rights, and protection against land grabs. Community-based land management approaches can also help ensure sustainable and inclusive land use.

2. **Protecting Seed Rights**

   Policies should protect farmers' rights to save, use, exchange, and sell farm-saved seeds. Intellectual property laws should balance the interests of farmers and seed companies, preventing monopolies and ensuring access to diverse and locally adapted seed varieties. Support for community seed banks and participatory breeding programs can enhance seed sovereignty.

3. **Enhancing Access to Resources**

   Governments and development agencies should invest in infrastructure, research, and extension services to improve farmers' access to resources. Financial support through grants, subsidies, and affordable credit can help farmers adopt sustainable practices and technologies. Promoting agroecological approaches can enhance resource efficiency and resilience.

4. **Ensuring Fair Market Conditions**

   Market policies should promote transparency, fair pricing, and access to markets for smallholder farmers. Support for cooperatives, direct marketing channels, and fair trade certification can improve farmers' incomes and bargaining power. Infrastructure development, such as roads, storage facilities, and digital platforms, can enhance market access and reduce post-harvest losses.

5. **Promoting Climate Resilience**

   Climate policies should support farmers in adapting to climate change through climate-smart agriculture, early warning systems, and disaster risk reduction strategies. Investments in research and extension services can help farmers adopt resilient practices and diversify their livelihoods. Social protection measures, such as crop insurance and safety nets, can provide security against climate shocks.

6. **Addressing Social and Gender Inequities**

   Policies should promote gender equality and social inclusion in agriculture. This includes ensuring equal access to land, resources, and services for women and marginalized groups. Gender-sensitive extension services, capacity-building programs, and support for women's cooperatives can enhance women's participation and leadership in agriculture.

Conclusion

Farmers' rights are fundamental to achieving sustainable agriculture, food security, and social justice. Advocacy for these rights requires a multifaceted approach, involving grassroots mobilization, legal and policy reforms, capacity building, and strategic alliances. By recognizing and supporting farmers' contributions, we can create enabling environments that empower agricultural communities and promote equitable and sustainable development. Collective action, informed by the experiences and knowledge of farmers, is essential to drive meaningful change and ensure that farmers' rights are upheld and respected globally. Through concerted efforts, we can build a more resilient and just agricultural system that benefits all.

References

Borowiak, C. (2004). Farmers' rights: Intellectual property regimes and the struggle over seeds. Politics & Society, 32(4), 511-543.

Carpenter, S. (2017). Family farm advocacy and rebellious lawyering. Clinical L. Rev., 24, 79.

Holt-Giménez, E., Bunch, R., Irán Vasquez, J., Wilson, J., Pimbert, M. P., Boukary, B., & Kneen, C. (2010). Linking farmers' movements for advocacy and practice. The Journal of Peasant Studies, 37(1), 203-236.

Shawki, N. (2014). New rights advocacy and the human rights of peasants: La Via Campesina and the evolution of new human rights norms. Journal of Human Rights Practice, 6(2), 306-326.

Ton, G., & Proctor, F. (2013). Empowering smallholder farmers in markets. Experiences with farmer-led research for advocacy. LEI.

Wells-Dang, A. (2013, April). Promoting land rights in Vietnam: A multi-sector advocacy coalition approach. In Annual World Bank Conference on Land and Poverty. Washington DC.

Winge, T., Adhikari, K., & Andersen, R. (2013). Advocacy for farmers' rights in Nepal. In Realising Farmers' Rights to Crop Genetic Resources (pp. 181-194). Routledge.

24

Agricultural Innovations from Around the World

Abstract

Agricultural innovations are crucial for addressing the pressing challenges of food security, environmental sustainability, and economic development. Around the world, innovative practices and technologies are transforming agriculture by increasing productivity, enhancing resilience to climate change, and improving livelihoods. This article provides an in-depth exploration of notable agricultural innovations, spanning traditional practices, cutting-edge technologies, and holistic approaches. By examining these innovations and their impacts, we can gain valuable insights into how agriculture is evolving to meet global needs.

Introduction

Agriculture is undergoing a profound transformation driven by technological advancements, sustainable practices, and innovative approaches. Innovations in farming have the potential to enhance productivity, resilience, and sustainability, thereby addressing global challenges such as population growth, climate change, and food insecurity. This article examines pioneering agricultural innovations from different continents, highlighting their applications, benefits, and implications for the future.

Innovations in North America

1 Precision Agriculture

Precision agriculture utilizes technologies such as GPS, sensors, and drones to optimize farm management practices. Farmers can apply inputs like fertilizers and water more precisely, reducing waste and improving yields. Case studies from the United States and Canada demonstrate the efficacy of precision agriculture in enhancing productivity while minimizing environmental impact.

2. Vertical Farming

Urban areas in North America are increasingly adopting vertical farming techniques to grow crops indoors, using stacked layers under controlled environments. This innovation conserves space, reduces water usage, and allows year-round production of fresh produce. Examples from cities like New York and Toronto illustrate the potential of vertical farming to supply local markets sustainably.

3. **Agricultural Biotechnology**

   Genetically modified organisms (GMOs) and gene editing technologies are revolutionizing crop breeding and pest management in North America. Biotechnological innovations have produced crops with enhanced resistance to diseases, pests, and environmental stressors. Case studies from the United States highlight the adoption and benefits of biotech crops in improving agricultural productivity and food security.

Innovations in Europe

1. **Agroecology and Organic Farming**

   Europe has been at the forefront of promoting agroecological and organic farming practices, emphasizing biodiversity conservation and sustainable resource management. Innovations include crop rotations, cover cropping, and integrated pest management techniques that reduce reliance on synthetic inputs. Examples from countries like France and Germany showcase successful transitions to sustainable agriculture and their positive environmental impacts.

2. **Smart Farming and IoT Applications**

   European farmers are increasingly adopting smart farming technologies and Internet of Things (IoT) applications to monitor crops, soil conditions, and livestock health in real-time. These technologies enable data-driven decision-making, optimizing resource use and reducing environmental footprint. Case studies from the Netherlands and Denmark demonstrate how IoT innovations are enhancing farm efficiency and sustainability.

3. **Urban Agriculture Initiatives**

   Cities across Europe are promoting urban agriculture initiatives to improve food security and sustainability. Rooftop gardens, community-supported agriculture, and hydroponic systems are transforming urban landscapes, providing fresh produce to local residents. Examples from cities like London and Berlin highlight the social, economic, and environmental benefits of integrating agriculture into urban planning.

Innovations in Asia

1. **Rice Intensification Techniques**

   In Asia, innovative rice cultivation techniques such as System of Rice Intensification (SRI) have increased yields while reducing water usage and greenhouse gas emissions. SRI emphasizes practices like transplanting young seedlings, reduced water application, and organic soil amendments. Case studies from India, China, and Vietnam illustrate the adoption and impact of SRI on smallholder farmers and rice production systems.

2. **Aquaponics and Integrated Farming Systems**

   Aquaponics, combining aquaculture and hydroponics, is gaining popularity in Asia for sustainable food production. Integrated farming systems that integrate livestock, fish, and crops maximize resource efficiency and diversify income streams for farmers. Examples from Thailand and Malaysia demonstrate the potential of aquaponics and integrated farming to enhance food security and rural livelihoods.

3. **Agricultural Robotics and AI**

   Asian countries are investing in agricultural robotics and artificial intelligence (AI) to automate farm operations and enhance productivity. Robots for planting, harvesting, and weeding tasks are revolutionizing labor-intensive agriculture, addressing labor shortages and improving efficiency. Case studies from Japan and South Korea showcase innovations in agricultural robotics and AI applications in precision farming.

Innovations in Africa

1. **Mobile Technology for Agriculture**

   Africa is leveraging mobile technology to provide farmers with access to market information, weather forecasts, and agronomic advice. Mobile apps and SMS-based platforms connect smallholder farmers to extension services and market opportunities, improving their decision-making and productivity. Examples from Kenya, Ghana, and Nigeria highlight the transformative impact of mobile technology on agricultural development.

2. **Drip Irrigation and Water Management**

   In arid and semi-arid regions of Africa, drip irrigation systems and water management innovations are crucial for sustainable agriculture. These technologies conserve water, improve crop yields, and enhance resilience to climate variability. Case studies from Ethiopia, South Africa, and Morocco demonstrate successful implementations of drip irrigation and water-efficient farming practices.

3. **Climate-Smart Agriculture**

   Climate-smart agriculture initiatives in Africa promote practices that increase resilience to climate change while reducing greenhouse gas emissions. Innovations include agroforestry, conservation agriculture, and climate-resilient crop varieties. Examples from Uganda, Senegal, and Tanzania illustrate how climate-smart agriculture can improve food security and mitigate the impacts of climate change on smallholder farmers.

Innovations in South America

1. **Agroforestry and Sustainable Land Use**

   South America embraces agroforestry systems that integrate trees with crops and livestock, enhancing biodiversity and ecosystem services. These sustainable land use practices improve soil fertility, water retention, and carbon sequestration. Case studies from Brazil and Colombia showcase successful agroforestry models and their contributions to environmental conservation and rural development.

2. **Bioclimatic Greenhouses**

   Bioclimatic greenhouses in South America use natural ventilation, solar energy, and climate control technologies to optimize growing conditions for crops. These innovative structures extend the growing season, increase yields, and reduce energy consumption compared to traditional greenhouse designs. Examples from Argentina and Chile demonstrate the benefits of bioclimatic greenhouses for sustainable horticulture and vegetable production.

3. **Organic and Fair Trade Certification**

   South American countries are leaders in organic and fair trade certification schemes, ensuring sustainable practices and equitable trade relations. Organic farming methods promote soil health, biodiversity, and consumer trust in food safety. Case studies from Peru and Ecuador highlight the economic opportunities and social benefits of organic and fair trade agriculture for small-scale farmers and indigenous communities.

Conclusion

Agricultural innovations from around the world are shaping the future of farming by improving productivity, sustainability, and resilience to global challenges. By learning from diverse experiences and adopting best practices, countries can accelerate progress towards achieving food security, reducing poverty, and promoting environmental stewardship. Continued investment in research, technology transfer, and policy support is essential to harnessing the full potential of agricultural innovation for the benefit of present and future generations.

References

Alston, J. M., & Pardey, P. G. (2021). The economics of agricultural innovation. Handbook of agricultural economics, 5, 3895-3980.

Bennett, D. J., & Jennings, R. C. (Eds.). (2013). Successful agricultural innovation in emerging economies: new genetic technologies for global food production. Cambridge University Press

Blakeney, M. (2022). Agricultural innovation and sustainable development. Sustainability, 14(5), 2698.

Delmer, D. P. (2005). Agriculture in the developing world: connecting innovations in plant research to downstream applications. Proceedings of the National Academy of Sciences, 102(44), 15739-15746.

Juma, C. (2015). The new harvest: agricultural innovation in Africa. Oxford University Press.

Sayer, J., & Cassman, K. G. (2013). Agricultural innovation to protect the environment. Proceedings of the National Academy of Sciences, 110(21), 8345-8348.